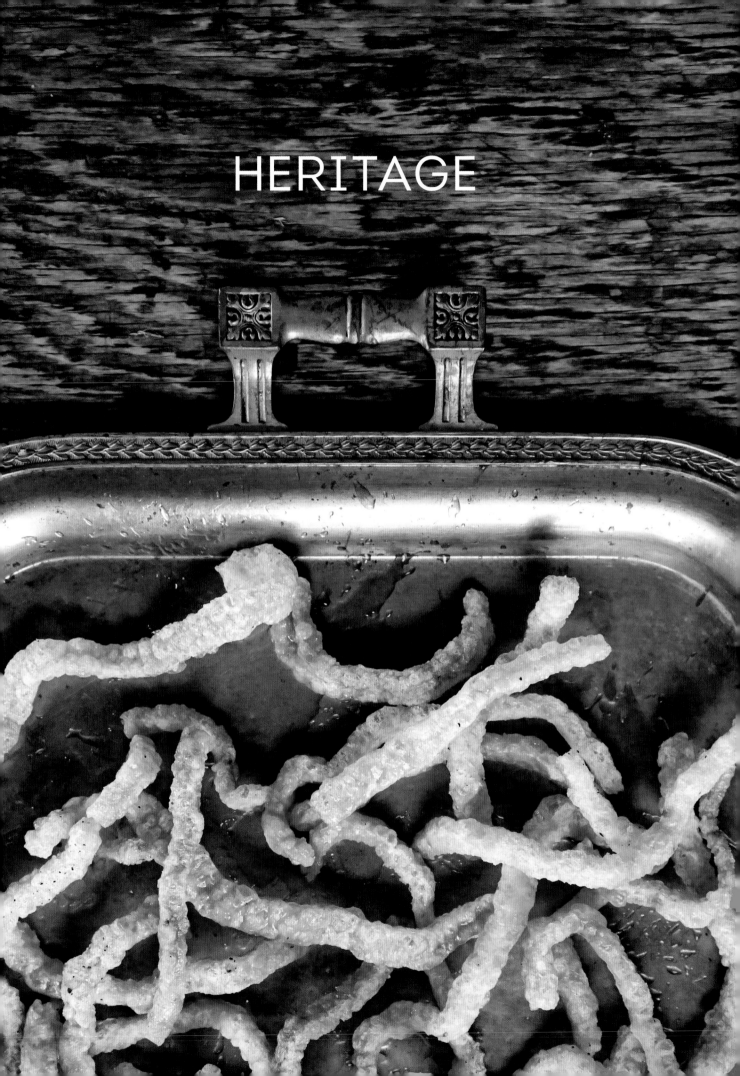

HERITAGE

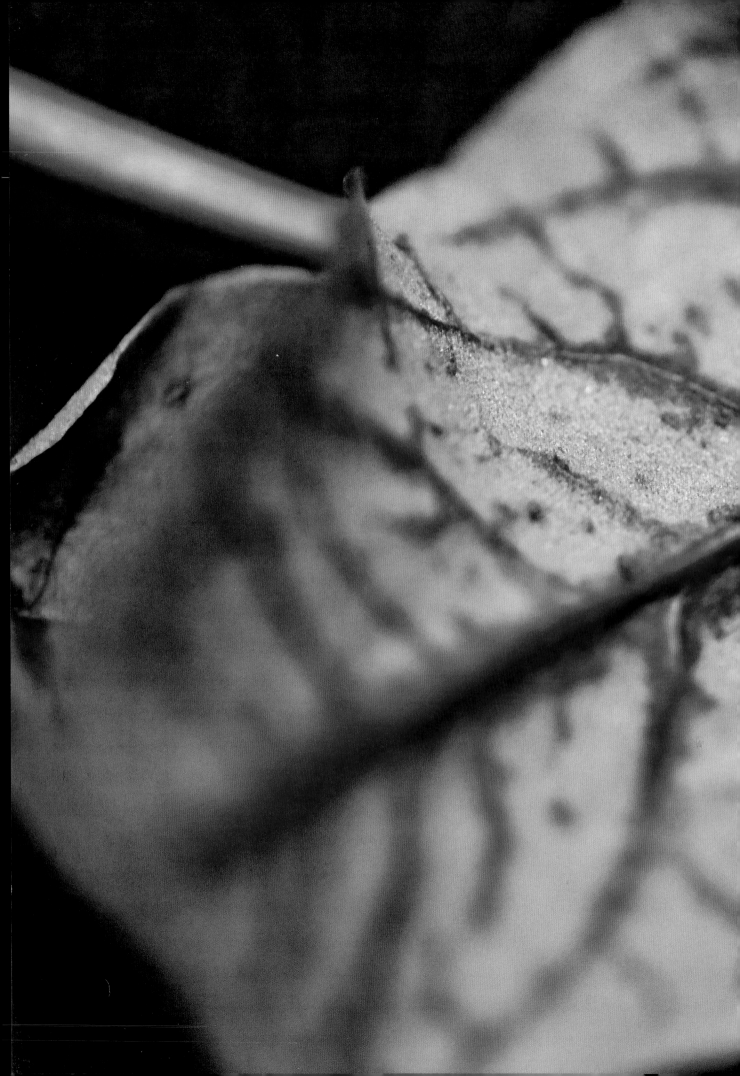

SEAN BROCK

HERITAGE

With contributions by
MARION SULLIVAN and JEFF ALLEN
Photographs by
PETER FRANK EDWARDS

ARTISAN

NEW YORK

To everyone who has believed in me, worked by my side, and eaten my food, especially Renee Brock, Homer Brock, and Audrey Morgan, for eternally inspiring me to work harder than the person beside me.

CONTENTS

I often tell people that I'm from the part of western Virginia that should have been in Kentucky. I grew up in Wise County, deep in the coalfields and hollers of the Appalachian Mountains. Have you ever seen the movie *Coal Miner's Daughter*? Well, that's what it looks like where I am from. That part of Virginia has a unique voice; you can hear it in the food, the art, the music, and the storytelling.

People in Southwest Virginia have a very distinct way of living. Most of them live off the land. If you don't have a kitchen garden, you're considered lazy. And if you don't have a freezer full of venison or catfish you caught yourself, you're not upholding the traditions passed down from your dad and grandpa. When I was a kid, everyone had a garden, and I mean everyone. It denoted status. Who could grow the best beans or the best tomatoes? Who had fewer weeds in her garden? Who took better care of his tractor?

If you grow up like I did, you learn to appreciate food on a different level. You see firsthand the work that goes into getting food onto the dinner table. You watch your family handle food with care and respect. It's in my blood; it's part of my DNA. My family loves food and appreciates it more than I can describe. Few people get as excited over a ripe tomato with salt on it or a perfectly executed cheeseburger from a diner as my mother does. I was taught to appreciate things that were made with care at a young age. It doesn't matter if it's chicken and dumplings or "Oysters and Pearls" from The French Laundry. If it's made with care, it is special.

I suppose it was my destiny to take over my father's coal trucking business and keep the Brock Trucking Company legacy alive. I would have been content surrounded by coal, literally—in my childhood memories there are piles of it as far as the eye can see, and I spent my days exploring old abandoned coal mines on my four-wheeler. I watched my father manage

a rowdy crew of bearded coal truck drivers with names like "Rooster Daddy" and "Fatboy." Coal provided for my family and all the families around me.

When I was eleven years old, my father died of a heart attack at the age of thirty-nine, and Brock Trucking was no more. My mother, brother, and I moved in with my grandparents, and those years would prove to be the foundation for my love affair with food. My grandparents had an enormous garden (much larger than what I see people calling farms these days), and it was their passion project. They plowed the fields with Haflinger horses and grew plants that were indigenous to our culture. My chores were those of a farmhand. While most kids would have hated these chores, I embraced them and actually looked forward to them. I loved being in the dirt and soaking up the sun. I loved the silence of the field, occasionally broken by a rooster's crow or a hungry horse's nicker. My grandmother Audrey was a master of many crafts. If she wasn't in the kitchen, she was in the garden, and her understanding of both domains was truly a marvel. My formative years with her were filled with amazement and respect. I was a very inquisitive child, and I asked too many questions. I wanted to know how things worked. I was lucky enough to have a grandmother who had most of the answers. I absorbed every piece of knowledge that she passed to me, and I wanted to be just like her. You could see the wisdom in her eyes, and you could see the years of work worn into her hands.

My grandmother's home was a beautifully mysterious playground, filled with bubbling vats of homemade wine and fermented ears of corn. Every square inch of her basement was covered with preserved food—I will never forget the smell of that basement. She saw that I loved food early on, and when I was thirteen or so, she bought me a hand-hammered wok from an infomercial. It came with a video and a cookbook. I watched the videos of workers forming the cast-iron woks with hammers at lightning speed. I was amazed that those hammer indentions served the purpose of holding the food up on the sides while the juices could run down and be reduced or thickened. I had grown up making biscuits and gravy, but this wok was something that opened up another part of the world to me. I started watching TV anytime there was a cooking show on. And just like that, I was hooked. I blame that damn wok for my food obsession. To this day, every time I use it I can hear my grandmother preaching about the importance of taking care of cast iron. I can only hope my wok lasts as long as the cornbread pans her grandmother gave her have.

You'll read more about my grandmother in the pages of this book, because she's been the greatest influence in my life. When I was a kid, we ate three meals a day at home. I thought that's what everyone else in America did too. In fact, I don't have a single memory of dining at a restaurant with my grandmother, and I probably didn't eat in a real restaurant until I was sixteen, which is pretty crazy considering what I do for a living. There were no restaurants in the town I lived in and only one sort of crummy grocery store. So you cooked what you grew, and you always knew where your food came from. That mentality influences everything I cook today, even if I no longer live in the Appalachians.

These days I am lucky enough to enjoy both the sophisticated foods that challenge me professionally and the comfort foods that nurture me on a regular basis. I appreciate and crave the best versions of both things. It's what drives my cooking. And so you'll find recipes for all kinds of foods—highbrow, lowbrow, and everything in between—within these pages. Some of the best food in the world is cooked in the most unassuming places by people who will never have their names in lights. And while I love caviar and foie gras dearly, I crave nostalgia. If you were to ask me what the most pleasurable thing in the world for me to eat is, I would tell you the story of Robo's Drive-In, located in my hometown of Pound, Virginia, and the amazing burgers they turn out daily for hungry truck drivers and peewee baseball stars. Nothing can beat eating a burger at a place where my dad took my mother on their first date.

Eventually we moved to the bigger town of Abingdon, and I got my first job as soon as I turned fifteen years and eight months (the legal age in Virginia at the time). I started out in the restaurant industry the way most people do: washing dishes. But I would study the cooks and their every move. I knew for sure that cooking was what I was meant to do. By the time I was seventeen, I had climbed the ranks and was working alongside the veterans I admired. I was head over heels in love with feeding people every day, and with being "in the weeds" in the kitchen. I knew that I had to go to culinary school.

My obsession with Lowcountry cooking started when I attended Johnson and Wales in Charleston, South Carolina. The food in Charleston was different from the food that graced my grandmother's table, but in some way I still felt connected to it. Southern food has enough soul to transcend region. I worked my butt off in some of the best kitchens in Charleston, learning how to make red rice, shrimp and grits, and hushpuppies. It didn't take me long to realize that the Lowcountry would become my culinary home. I cut my first teeth on the produce in my grandmother's rural Virginia garden, but the Lowcountry shaped the cook that I was to become.

In early 2003, at the age of twenty-four, I took my first chef job, at the Hermitage in Nashville. It would prove to be the most difficult few years of my life. I surely wasn't ready to helm the kitchen of a five-star hotel, but I learned a lot about myself during that time and a lot about the industry. It was worlds apart from my days as a line cook and I was in over my head, yet the move allowed me to discover a city that spoke to me. Nashville is full of amazing music and delicious food traditions. While I knew in the back of my mind that Charleston would eventually become my home, I also knew that I wanted to return to Nashville someday. It was just too much fun to leave behind for good.

When I accepted the chef position at McCrady's in 2006, my dream of running a kitchen in Charleston finally became a reality. I could not have been more nervous or more grateful. The Lowcountry is a diverse region filled with a heritage of deeply rooted traditions. It is a landscape of extraordinary beauty. The foodways here are old and elemental but speak with the authority of a hard-fought past. The people in Charleston deeply appreciate their heritage foods made with local ingredients, and they respect the people who still cook them. The ingredients come from people who revere them, and the methods are as sacrosanct as the ingredients. This food represents the living history of the Lowcountry, and I have always wanted to do my best to honor that.

It is these life lessons that have influenced my career. It wasn't until I moved back to Charleston in my late twenties that I fully realized how important my Appalachian upbringing was. I started to spend time with the people who work the fields and waterways of the Lowcountry, and I was struck by their similar way of thinking. They live off the land the old-fashioned way: corn is saved and passed down because of the quality moonshine and grits it produces; red peas isolated for years on a barrier island are prized, because that is what generations before grew and cooked; the art of throwing a cast net is a tradition passed down to every Lowcountry boy from his grandfather. That two regions so vastly different could be so profoundly connected through a shared way of thinking was an epiphany. It took coming to Charleston to teach me that the soul of Southern food knows no geographical boundaries.

I believe that this kind of revelation can happen for anyone, whether it's in your own backyard or somewhere in your travels. I've learned that it's important to appreciate and respect where you come from. That sense of pride has guided me, and I hope to inspire you to understand your own heritage.

THE STORY OF HOPPIN' JOHN

A man once brought some Sea Island "red peas" to McCrady's and told me they were special. They were red cowpeas that had been planted originally by African slaves in the Sea Islands. He also gave me a bag of Carolina Gold rice, which he was trying to reintroduce to the area. These are the ingredients of hoppin' John, the quintessential Lowcountry dish.

I wanted to believe him, but I hadn't had the best experience with hoppin' John. The first time I sank my teeth into a bowl of rice and peas, I was eighteen years old and I was excited to experience the dish I had read so much about. I had painted the most romantic picture possible in my head, and I couldn't have been more let down. My first bite was beyond disappointing. So I took another bite . . . more of the same. I left that meal wondering what all the fuss was about—it may have been one of the blandest things I had ever tasted. Little did I know I was being taught the lesson of my life. The lesson that would change the way I looked at food for the rest of my career. That first bowl of hoppin' John was so tasteless that I lost interest in Lowcountry cooking during my time at cooking school.

It wasn't until I tasted those Sea Island red peas in a bowl together with Carolina Gold rice that I realized what had gone wrong those many years ago when I'd first tasted hoppin' John. That hoppin' John was made with commercial, enriched rice and old, flavorless black-eyed peas. At that time the heirloom products that had helped shape the culture of Charleston weren't available to a chef for purchase. They simply weren't being grown. After the Civil War and the abolition of slavery, rice in Charleston was no more. The last commercial harvest was sold in 1927. What emerged after the Great Depression was a modified commercialized rice brand, with a very different flavor and texture from those of the rice people remembered from their youth.

Thanks to the work of the Carolina Gold Rice Foundation (founded by Merle Shepard, David Shields, Glenn Roberts, and other like-minded individuals), I could finally understand why hoppin' John is such a celebrated dish. They did the research and tested old samples of rice from the Carolina Gold Rice era. They collected heirloom seeds and grew heirloom rice, along with benne, okra, peanuts, sorghum, and cowpeas to rotate the crops, and each of these ingredients adds its own particular magic to the soil in which the rice grows. What resulted was the most flavorful rice I have ever tasted.

The foundation's quest is a never-ending one, but all these products are now available commercially through Anson Mills, a company started by Glenn Roberts to repatriate the Carolina Rice Kitchen, so chefs and everyone else can sample these heirloom ingredients (see Resources, page 326). Make this dish (see page 16) with rice and red peas from Anson Mills and taste the difference. Real hoppin' John lies at the soul of the Lowcountry—a metaphor of its history and culture. It embodies the marriage between the golden rice seed, which crossed the Atlantic to underwrite the elaborate wealth of Charleston, and the lowly cowpea, a West African native originally deemed fodder for cattle and for the slaves who had brought that rice to Carolina and grew it. The cultures and ingredients coalesce in this marvelous dish.

LOWCOUNTRY HOPPIN' JOHN
SERVES 6 TO 8

PEAS

2 quarts Pork Stock (page 319) or Chicken Stock (page 318)

1 cup Anson Mills Sea Island Red Peas (see Resources, page 326), soaked in a pot of water in the refrigerator overnight

1½ cups medium dice onions

1 cup medium dice peeled carrots

1½ cups medium dice celery

2 garlic cloves, thinly sliced

1 fresh bay leaf

10 thyme sprigs

½ jalapeño, chopped

Kosher salt

RICE

4 cups water

1 teaspoon kosher salt

¼ teaspoon cayenne pepper

1 cup Anson Mills Carolina Gold Rice (see Resources, page 326)

4 tablespoons unsalted butter, cubed

RED PEA GRAVY

Reserved 1 cup cooked red peas

Reserved 2 cups cooking liquid from the peas

1 tablespoon unsalted butter

Cider vinegar

Sliced chives or scallions for garnish

FOR THE PEAS: Bring the stock to a simmer in a small pot. Drain the peas and add to the stock, along with all of the remaining ingredients except the salt. Cook the peas, partially covered, over low heat until they are soft, about 1 hour. Season to taste with salt. *(The peas can be cooked ahead and refrigerated in their liquid for up to 3 days; reheat, covered, over low heat before proceeding.)*

Drain the peas, reserving their cooking liquid, and measure out 1 cup peas and 2 cups liquid for the gravy; return the rest of the peas and liquid to the pot and keep warm.

MEANWHILE, FOR THE RICE: About 45 minutes before the peas are cooked, preheat the oven to 300°F.

Bring the water, salt, and cayenne pepper to a boil in a large saucepan over medium-high heat. Reduce the heat to medium, add the rice, stir once, and bring to a simmer. Simmer gently, uncovered, stirring occasionally, until the rice is al dente, about 15 minutes.

Drain the rice in a sieve and rinse under cold water. Spread the rice out on a rimmed baking sheet. Dry the rice in the oven, stirring occasionally, for 10 minutes. Scatter the butter evenly over the rice and continue to dry it, stirring every few minutes, for about 5 minutes longer. All excess moisture should have evaporated and the grains should be dry and separate.

FOR THE GRAVY: Put the 1 cup peas, 2 cups cooking liquid, and the butter in a blender and blend on high until smooth, about 3 minutes. Add cider vinegar to taste.

(The gravy can be made up to 3 days ahead and kept in a covered container in the refrigerator; reheat, covered, over the lowest possible heat, stirring occasionally to prevent scorching.)

TO COMPLETE: Use a slotted spoon to transfer the peas to a large serving bowl. Add the rice and carefully toss the rice and peas together. Pour the gravy over them, sprinkle with chives or scallions, and serve.

ELEVATING SOUTHERN CUISINE

The older I get and the more I travel, the more I realize what an ongoing challenge it is to master my craft. I've been lucky to get to travel the world studying food. It has taught me so much— whether it's the incomparable respect for ingredients chefs have in Japan, or the way smoked and fermented seafood is used in dishes in West Africa. With every new place that I visit, I pick up something along the way that I can apply to my cooking.

One of the things that I've learned is how misunderstood Southern cuisine is—it is a lot more complicated than most people realize. You could spend your entire life studying the cuisines and cultures of the American South. It's about as diverse and as large as Europe! As I experience other cultures and traditions, I've realized how precious the South's ingredients and traditions are.

With the opening of Husk, I have been fortunate to be able to make a small difference in the way many people view Southern food. I decided that Husk could be an amazing platform for studying and celebrating the American South. I knew the truth

and beauty in our cuisine, and I wanted to share it with the world. In order for Husk to be a truly Southern restaurant, it was essential that our guests experience the South's unique *terroir* (meaning the characteristics—including the people, places, and methods used to grow an ingredient—of a locale and its food). So I decided that we would purchase only ingredients sourced south of the Mason-Dixon Line. That limitation has allowed me to grow as a cook at a much more rapid pace than I could have ever imagined.

Cooking without ingredients like olive oil and vinegar was an eye-opener for me. But that challenge became a major learning experience, and now we make more than fifty of our own vinegars, all Southern flavor based, with things like sorghum and muscadine. We grind our own flour, and we found a Southern producer of olive oil.

Produce availability was a challenge at first too: we are limited to using only what is made or grown within a thirty- to forty-mile radius. That means, for example, we can use only Florida citrus or the small amount of citrus grown in Charleston. And citrus is in season only in the winter. So we have to be creative in those months. But these limitations forced us to plan ahead and preserve for the off seasons. We pickle, ferment, and can throughout the summer and fall, an essential part of keeping the restaurant running year-round. At first, determining the amounts we needed was tricky, and working with crops made us face realities that farmers have dealt with for the ages. The first year, we planted an entire field of potatoes with the anticipation of using them the following season. The crop failed, we were out thousands of dollars, and we had no potatoes. The incident connected me to our farmers in a way that I never expected. Now I look at these challenges as opportunities to deepen my understanding of how food systems work.

Husk Nashville has the same mission, but with a different *terroir.* We have very little seafood in Nashville, so the menu is filled with pork, beef, and an array of vegetables. And there are four true seasons in Nashville, so the produce is different from what we get in Charleston. The seafood available in Nashville is farm-raised seafood, and I'm using ingredients like trout and catfish, which are what I ate growing up in the Appalachian Mountains. Winters in Nashville remain a challenge for us, so building our larder is as important as always. We have more corn in Nashville and much less rice than in Charleston. This all underscores how different the South is regionally. And these microregions create such diverging cuisines. The challenges are different, but the work is the same. Opening Husk in Nashville, of course, also brought me back to another city I hold dear.

ABOUT THE BOOK

I've always collected cookbooks on my favorite subjects, but for the past few years I have been attempting to get copies of all of the cookbooks published in America in the nineteenth century. Sitting around and flipping through these old books and soaking up the knowledge of that era, I realized how important it is to document such things, and how inspiring those old recipes were to me. Were they perfect recipes? Did they work every time? Were they insanely delicious? The answer is, not always. But they taught me a much more important lesson than what ended up in the pot or on the plate—to be more observant, to capture the moment, learn from it, and respect it. So I decided to write this book in the hope that it will inspire people to take the same journey that I have: to embrace their heritage and, most important, to celebrate what they find and share it with others.

I can promise you that I have made these recipes hundreds of times. My team and I tested them and retested them, and we have served most of them at the restaurants to our guests. But I also know that these dishes are going to taste different when you cook them. That's just the way it is. Every batch of carrots we get through our back door tastes a little different (the carrots you get will surely taste different). And so depending on the ingredients, we adjust the seasonings to keep the dishes balanced, not just on a daily basis, but on a pan-to-pan basis during service. And, of course, we are using products that are grown in the South. Therefore, if you live elsewhere, the *terroir* of your products is going to be different. So, what does all this mean? It means that I want you to find and gather ingredients in your own region and make these recipes your own.

The ingredients that thrive in your part of the world are the ingredients that you should seek out and make the focus of your cuisine. The rest will fall into place. When you buy local and fresh, the ingredients will be so good that you won't have to do much with them for them to taste good. Those ingredients also carry a historical significance, which can prove to be a source of inspiration when you're cooking and a great source of pride within the community. Do your own research and find the foods that make your region unique. The mix of culinary influences may turn into a revelation and change the way you cook and eat. The Lowcountry dishes that you find throughout this book are themselves a confluence of African, European, and American Indian influence, thrown together long ago in the mixing bowl of history and refined by generations of grandmothers. I've only added my own vision and passed them along to you.

The recipes in this book range from simple dishes requiring little culinary skill to composed ones with complex techniques requiring the fancy equipment we have at the restaurants. The recipes are organized into chapters categorized by where the food comes from. So "The Yard," for example, includes the simple skillet-roasted chicken that I cook at home, a version of the fried chicken I cook at Husk, and chicken and fowl recipes that you might find on the menu at McCrady's. If you're craving chicken, I suggest you start by cooking the Chicken Simply Roasted in a Skillet (page 109); it's easy and insanely delicious. Then this might inspire you to work your way up to the Roasted Duck (page 116), or build a pit in your backyard to roast a pig (see page 140), or to try your hand at some of my grandmother's desserts, like her Apple-Sorghum Stack Cake (page 284). At the end of the day, I just want to inspire as many people as possible to cook with care and passion.

When I cook, I take notes and document my trials and tribulations as well as my successes, always trying to push a recipe forward. My grandmother did the same on index cards that overflowed from a wooden box on the kitchen counter and on the labels that were pasted on dusty Mason jars in the cellar. Notebooks help—trust me; I'd be lost without mine. I have saved mine from all the way back to my job at Peninsula Grill in 1998. I really enjoy looking back and seeing how far my cooking has come and how my point of view has changed. Most, if not all, of the recipes in this book come from these old notebooks. So I encourage you to record your own experiences when you are cooking. Track what worked and what didn't and use your notes as a learning tool. They'll help you to improvise and make the recipes your own.

I believe that both professional chefs and home cooks can move their cuisine forward by understanding the past and knowing where their food comes from. I hope that my experiences will move more people to research their heritage and find inspiration from the food and traditions they grew up with. I hope that a few folks will put this book down and plant a garden, even if it's just a little herb garden outside the back door. I also hope they choose plants that reflect the heritage of the place they live.

Writing this book, I found inspiration in the South Carolina Lowcountry, and I've shared the stories of a number of heroes of mine in these pages. But, ultimately, it's a book about the South and all the things that helped shape a culture. It's about cooking, storytelling, agriculture, music, landscapes, history, folk art, writing, and everything else that is right underneath my boots. Often, the truest form of inspiration is right where you are standing. This book is my attempt to capture that spirit through cooking and sharing. A spirit that celebrates traditions past while moving forward with an earnest goodwill.

I'll never claim to be an expert on Lowcountry cooking. Hell, I grew up in the coalfields of southwest Virginia—what do I know about she-crab soup? My point is, find something you love, something that speaks to you, and embrace it. Through my journey trying to understand the cuisine of the rice kitchen, my eyes were opened to how diverse the South truly is. It instilled in me a deep respect for each of its unique regions. I could spend my entire career studying Southern food and still never fully understand it, but I'll never stop trying. At the end of the day, it's not about becoming an expert, it's about the lessons learned along the way. And it's about fulfilling my mission to reinvent Southern cooking as one of the great cuisines of the world.

MY MANIFESTO

- Cook with soul—but first, get to know your soul.

- Be proud of your roots, be proud of your home, be proud of your family and its culture. That's your inspiration.

- Cook as if every day you were cooking for your grandmother. If your grandmother is still alive, cook with her as much as possible, and write everything down.

- Respect ingredients and the people who produce them.

- Visit the farmers' market at least once a week, and use most of your food budget at the market.

- Buy the best that you can afford.

- Grow your own—even if it's just a rosemary bush. You'll taste the difference and start planting more right away.

- Do as little as possible to an ingredient when it's perfect and at its peak.

- You can never be too organized; a clean work space allows for a clean mind that can produce a clean plate of flavors.

- Cook in the moment. Cook the way you are feeling, cook to suit the weather, cook with your mood, or to change your mood.

- Let vegetables tell you what to do. Taste them raw before you start thinking about how to cook them. Are they sweet? tender? crunchy? starchy?

- Cook a vegetarian feast occasionally. Vegetables cooked with care can be just as rewarding as a piece of braised meat.

- If you are dead set on making a specific recipe but when you go to the market the ingredients don't speak to you or feel and smell perfect, don't make the recipe. Cook from the hip—you may surprise yourself. Perfect ingredients don't require much; shop for flavor, not concept.

- Overseason something with salt and acid just so you know what is too much. Then ride the line, and you'll find your balance.

- Listen to your tongue; it's smart.

- Cook using your instincts. Cooking times are just guidelines.

- Try to make every dish better every time you make it. Keep a notebook to document successes and failures. And record your creative inspirations in it as well.

- Eat with your hands as much as possible.

- Be curious! Ask yourself questions: Why did the fish stick to the pan? Why did my sauce break?

- Never stop researching and seeking knowledge in the kitchen.

- Cooking should make you happy. If it starts making you angry, stop cooking and go eat at a nice restaurant. Come back the next night and think about what went wrong and give it another shot.

- He who dies with the biggest pantry wins.

THE GARDEN

IT BEGINS
IN THE DIRT

A garden is a magical place. I still remember my grandmother's: the long rows of potatoes, cabbage, tomatoes, and corn; the hot dusty soil; and the smell of summer rain. I grew up there, wandering through the warm dirt and reveling in intoxicating smells. I remember my initial amazement when I learned that potatoes came from under the ground. I spent the harvest carving into just-dug potatoes with a pocketknife, drinking up the truffled aroma of the moist tubers mixed with earth and eating them raw, straight from the field.

My mom tells everyone that I teethed on rhubarb, saving her a bunch of money on pacifiers, and I think it's fair to say that my experiences within the garden led to my becoming a chef. A vibrant garden is a social space, one that requires a lot of work, both in the dirt and after the harvest. My family used to sit out under an old walnut tree in the backyard, telling family stories as we shaved cabbage into a crock to make my grandmother's mixed pickles. It was a Southern version of kimchi, with a bit less spice but all of the same love. We'd work all summer to grow cabbage, corn, and snap beans, then salt them in crocks and store them in the basement to ferment; or we'd string up wild goose beans with a needle and thread and dry them into "leather britches" outside on the porch. The beans shrink up and become super-intense in flavor after they're dried, and when you cook them they're transformed—it's like eating meat. I can remember sitting around a big table as a kid, stringing them as a communal exercise. And although I hated the tedious work then of picking long rows and preparing the beans for cooking or canning, I now realize what a bonding experience it was for my family, and I am saddened that so many people have lost that common thread.

The garden is an integral part of McCrady's and Husk. Years ago, I experimented with leasing a plot of land on the barrier island of Wadmalaw that sits just south of Charleston, planting just about every seed that I could find. My first attempts were complete failures. I knew so little and I lost whole crops to disease (not to mention the digging of my pug, Yuzu).

Over time, I learned a lot about the land, its disease environment and climate, and the connection of those things to the way people eat. These are things that I could never learn in a restaurant kitchen. Instead of hanging out with other chefs, we seek out local farmers, and this brings our whole team closer to the food with which we work. The kitchen is transformed by an understanding of the function of the garden.

I take the staff with me to these gardens. Servers and kitchen staff at McCrady's have worked the fields, and together we have all gained a better understanding and appreciation of what it truly means to eat in Charleston. Your outlook changes when you grow something from seed to stalk. You create ownership and a sense of pride and accomplishment. You start to think more deeply about the process of food itself (I briefly toyed with preparing "raw foods," loath to manipulate such a wonderful thing as nature). I find that the connection between table and dirt runs through a farmer and his or her garden just as much as any chef's inspiration.

In many places across the country, our heritage is threatened. In the Lowcountry, crops like heirloom benne, Sea Island rice peas, and Choppee okra are almost impossible to find now unless you grow them yourself. And yet prior to the twentieth century, they were common staples, available to all. Over the last few decades, a transition to large-scale commercial agriculture has occurred, one that values disease resistance and plant yields over flavor and timeworn tradition. This has changed the way people eat, even in a place so steeped in history as Charleston.

So my aims are twofold: to help bring the small local farmer back to prominence by respecting the work of local growers and to encourage farmers to reach back beyond the hybrid varieties, wasteful practices, and chemical inputs that have transformed agriculture (and the taste of food) over the last century. Only by reclaiming the flavors unique to Charleston or Nashville, or any locale, can we begin to move forward. Otherwise, no one will even know what's missing, and we will have lost forever a tradition that transcends the mere practice of producing food. Authentic food must engage its geographic culture—it must reflect a way of life.

CORN-GOAT CHEESE SOUP
WITH SHRIMP AND BROWN BUTTER CHANTERELLES

SERVES 6

Real sweet corn is a short-lived pleasure, and I invariably find myself gorging on half a dozen ears at the start of every season. As fast as it comes in, it's gone, which only heightens the anticipation of the season. I think it appropriate to pair something as fleeting as this with an equally short-lived garnish, such as the local chanterelles that foragers bring by for just a few weeks each summer.

This simple soup includes shrimp and mushroom garnishes, but you can leave them out and still have a good bowl. You can also use this base for other dishes. Try swirling in a few tablespoons of butter and an extra ladleful of vegetable stock before napping it as a sauce over a piece of roast grouper, or maybe including a little less vegetable stock and placing the thicker puree underneath the fish.

At McCrady's, we also garnish the soup with freshly popped popcorn and tiny sprigs of fresh chervil.

NOTE

When cooking chanterelles, make sure that the liquid that they release has completely evaporated before adding any other liquid. Moisture content varies from batch to batch.

SOUP

4 cups Vegetable Stock (page 316)

4 large ears sweet corn

3 tablespoons unsalted butter

2 marjoram sprigs

3 ounces goat cheese

Kosher salt and freshly ground white pepper

SHRIMP

4 lemons

4 cups Vegetable Stock (page 316)

1 cup dry white wine

1 fresh bay leaf

1 tablespoon yellow mustard seeds

1 tablespoon whole black peppercorns

1 teaspoon celery seeds

12 large shrimp (16–20 count)

2 tablespoons crème fraîche

1 tablespoon chopped basil

Kosher salt

Cayenne pepper

CHANTERELLES

20 large chanterelles, stems scraped clean, tops swirled in warm water to clean, and dried

3 tablespoons unsalted butter

2 thyme sprigs

FOR THE CORN STOCK: Put the vegetable stock in a medium pot and set over medium heat. Cut the kernels from the corn; set aside. Using a box grater, scrape the "milk" from the cobs into the stock. Cut the cobs in half and add them to the stock. Bring to a simmer for 30 minutes. Strain the stock and set aside.

MEANWHILE, FOR THE SHRIMP: Cut 3 of the lemons in half. Put the lemon halves, vegetable stock, wine, bay leaf, mustard seeds, peppercorns, and celery seeds in a medium pot and bring to a boil over high heat. Reduce the heat to low and simmer for 30 minutes.

To devein the shrimp without removing the shells, take a paring knife and cut each one down the back, through the shell. You will see a vein, which is the shrimp's digestive tract; lift it out with the tip of the knife and discard.

After the stock has simmered for 30 minutes, remove the pot from the heat, add the shrimp to the stock, and cover the pot. After 7 minutes, transfer the shrimp to a colander to drain. Strain the shrimp stock. Reserve ¾ cup to braise the chanterelles and freeze the rest for making a soup or sauce.

When the shrimp are cool enough to handle, peel them and cut them into thin slices. Set aside. Zest and juice the remaining lemon. Set aside.

TO COMPLETE THE SOUP: Heat the butter in a medium pot over medium-high heat. When the butter is foamy, add the corn kernels and marjoram and cook, stirring, just until the corn is tender, about 3 minutes. Add 3 cups of the corn stock and bring to a simmer. Remove the marjoram and discard.

Working in batches, puree the soup in a blender until very smooth, about 5 minutes. Pour into a saucepan. The soup can be kept warm over low heat for up to 20 minutes.

FOR THE CHANTERELLES: Use your hands to gently pull the chanterelles lengthwise into quarters. Put a skillet large enough to hold the chanterelles in one layer over very high heat (work in batches). When the skillet is very hot, add the butter. As soon as the butter melts, add the mushrooms and thyme and sear the mushrooms, without moving them or shaking the skillet, until all liquid has evaporated and the butter has browned, about 7 minutes. Add just enough shrimp stock, about ½ cup, to create a glaze. Keep warm at the back of the stove.

TO COMPLETE: Return the soup to the blender, add the goat cheese, and blend until smooth, about 2 minutes. Season with salt and white pepper. Combine the lemon zest and juice, crème fraîche, basil, and sliced shrimp in a glass bowl and toss to coat the shrimp. Season with salt and cayenne pepper.

Divide the soup among six warm bowls. Place the shrimp and the chanterelles in the bowls. Serve immediately.

MUSCADINE-CUCUMBER GAZPACHO

SERVES 6

The inspiration for this dish was the classic Spanish white gazpacho, which often features grapes. I use the indigenous muscadine grape, a thick-skinned subgenus of some three hundred varieties that thrives in the heat and humidity of the South. Muscadine grapes have a unique flavor, musty and floral. They go well with the yeasty aromas of the bread and the fruitful zing of the sherry vinegar. Make sure to use good-quality vinegar and olive oil. You can taste the difference.

20 large muscadine grapes

6 pounds cucumbers, peeled, halved lengthwise, and seeds removed

½ cup extra-virgin olive oil

½ cup thinly sliced leek, white part only, washed and drained

6 slices white sandwich bread, crusts removed

½ cup Marcona almonds

1 tablespoon sour cream

Kosher salt

Sherry vinegar

EQUIPMENT

Juice extractor

Press the grapes through a fine-mesh sieve into a bowl to obtain the juice. Discard the skins, seeds, and pulp.

Dice about 3 pounds of the cucumbers; you want 4 cups dice. Run the remaining cucumbers through a juice extractor; you want 2 cups juice.

Heat 1 tablespoon of the olive oil in a medium skillet over medium-low heat. Add the leeks and cook, stirring frequently, until translucent and tender, about 5 minutes. Cool.

Working in batches if necessary, combine the muscadine juice, diced cucumbers, cucumber juice, the remaining 7 tablespoons olive oil, the leeks, bread, almonds, and sour cream in a blender and blend until very smooth, about 5 minutes. Season with salt.

Transfer to a container, cover, and refrigerate until well chilled. *(Tightly covered, the soup will keep for up to 3 days in the refrigerator.)*

When ready to serve, season the gazpacho with sherry vinegar.

CHILLED FENNEL BISQUE
WITH CITRUS-CURED SCALLOPS AND ALMOND OIL

SERVES 4

We make this soup with a simple braise of fennel combined with vegetable stock. It allows the purity of the fennel to shine through, because you're not letting any aromatics like garlic or the flavor of chicken stock get in the way. Make the soup with the first baby fennel bulbs of the season, when the freshness and sweet high notes fully sing. In a simple preparation like this one, quality matters even more than usual.

BISQUE

2 pounds baby fennel bulbs with tops

3 cups Vegetable Stock (page 316)

3 tablespoons unsalted butter

Kosher salt and freshly ground white pepper

4 ounces cream cheese

Almond oil

SCALLOPS

6 large (U-10) dry-packed sea scallops, tough side muscles removed, cut into ½-inch dice

Grated zest of 1 lime (use a Microplane)

Juice of 3 limes

1 tablespoon almond oil, plus more for drizzling

1 tablespoon chopped cilantro

1 teaspoon fennel pollen

Kosher salt and freshly ground white pepper

FOR THE BISQUE: Remove the stalks and fronds from the fennel; reserve 4 fronds for garnish. Cut the bulbs crosswise into ¼-inch-thick slices. Set aside.

Bring the vegetable stock to a simmer in a medium pot over medium heat. Add the fennel stalks and the remaining fronds and simmer until tender, about 30 minutes. Keep warm over low heat.

Heat the butter in a medium pot over high heat. When the butter is foamy, add the sliced fennel. Reduce the heat to low and braise the fennel, stirring occasionally, until very soft, about 8 minutes. Add the stock and bring to a simmer. Season with salt and white pepper.

Working in batches if necessary, transfer the soup, along with the cream cheese, to a blender and blend until very smooth, about 5 minutes. Strain the soup into a container and let cool to room temperature. Cover and refrigerate until chilled. *(Tightly covered, the soup will keep for up to 3 days in the refrigerator.)*

FOR THE SCALLOPS: Put the scallops in a chilled glass bowl. Add the lime zest and juice and stir to combine. Cover and refrigerate for 20 minutes.

Just before serving, add the almond oil, cilantro, and fennel pollen to the scallops. Season with salt and white pepper.

TO COMPLETE: Divide the soup and scallops among four bowls. Drizzle a little almond oil on top of the soup and garnish with the reserved fennel fronds.

GREEN GARLIC BISQUE
WITH HERBED BUTTERMILK AND FRIED GREEN TOMATO CROUTONS

SERVES 6

Green garlic is the business end of the immature scapes put out by the garlic plant before beginning bulb production, and it's one of my favorite ingredients. It only has about a six-week window in the South, so you have to act fast when you see it in the market. If you catch the garlic while it's about the size of a young leek and is still very soft, the flavor is subtle and dynamic.

In this recipe, herbed buttermilk provides just enough creaminess and acidity to balance the strong flavor of the garlic. The "croutons," with their own balance of crunchy goodness and green tomato acidity, add lots of texture and flavor. They can also be made and passed as a noteworthy canapé on their own, perhaps dipped in fresh herbed buttermilk just before eating.

HERBED BUTTERMILK
1 cup whole-milk buttermilk

1 tablespoon chopped flat-leaf parsley

1 tablespoon chopped tarragon

1 tablespoon minced chives

1 tablespoon chopped chervil

Kosher salt

Cayenne pepper

BISQUE
5 whole green garlic stalks (about 6 ounces), trimmed

3 tablespoons unsalted butter

4 cups Vegetable Stock (page 316)

1 cup heavy cream

¾ cup tarragon leaves

Kosher salt and freshly ground white pepper

FRIED GREEN TOMATO CROUTONS
2 cups rice wine vinegar

Kosher salt

1 tablespoon sugar

2 tablespoons chopped tarragon

5 large green tomatoes (about 2½ pounds), cored and cut into 1-inch cubes

2 quarts canola oil, for deep-frying

3 cups Anson Mills French Mediterranean White Bread Flour (see Resources, page 326)

4 cups whole-milk buttermilk

3 cups cornmeal, preferably Anson Mills Antebellum Coarse Yellow Cornmeal (see Resources, page 326)

Freshly ground black pepper

FOR THE HERBED BUTTERMILK: Stir the buttermilk and herbs together in a bowl and season with salt and cayenne. Cover tightly and refrigerate. *(The buttermilk can be made up to 3 days ahead.)*

FOR THE BISQUE: Separate the white parts from the green parts of the green garlic. Wash and dry, keeping the tops and bottoms separate. Chop the whites. Slice the greens.

Heat the butter in a medium pot over high heat. When the butter is foamy, add the white parts of the garlic, reduce the heat to medium-high, and cook until very soft, about 12 minutes; add a splash of the vegetable stock if the whites start to stick. Add the green parts of the garlic and the (remaining) stock, bring to a simmer, and cook for 10 minutes. Add the cream and bring back to a simmer.

Working in batches if necessary, transfer the soup, along with the tarragon leaves, to a blender and blend until very smooth, about 7 minutes. Season with salt and white pepper. Strain the soup into a saucepan and set aside.

FOR THE FRIED GREEN TOMATO CROUTONS: Put the vinegar, salt, and sugar in a large bowl and stir until the salt and sugar are dissolved. Stir in the tarragon, add the tomatoes, cover, and marinate at room temperature for 20 minutes.

Pour the canola oil into a deep fryer or a large pot and heat it to 350°F. Meanwhile, make a breading station by setting up three separate shallow bowls for the flour, buttermilk, and cornmeal.

Working in batches, dredge the tomato cubes in the flour, gently shaking off the excess, dip them into the buttermilk, allowing any excess to fall back into the bowl, and then dredge them in the cornmeal, gently shaking off any excess. Fry until crispy and golden brown, about 5 minutes. Drain on paper towels and immediately sprinkle lightly with salt and pepper.

TO COMPLETE: Divide the soup among six warm bowls. Garnish each serving with 5 green tomato croutons (extra croutons make a great snack). Finish with a generous drizzle of the herbed buttermilk.

SHAWN THACKERAY

Wadmalaw Island is not a normal place. About thirty minutes south of Charleston proper, it's a hidden Eden, at least for a country boy. There are no restaurants, no gas stations, and, in fact, no businesses at all. It's just beautiful farmland, and Shawn Thackeray's farm fits right into the landscape. Shawn is a custom tomato grower; he's been growing tomatoes since he was seventeen.

Shawn found an opportunity on Wadmalaw, because few farmers on the island were interested in specialty produce until recently. Most people there are busy mining the soil's fertility to raise corn and truck it away to distill into ethanol. But I'm convinced that specialty produce, grown locally and with care, can transform the way we eat, and this is where Shawn Thackeray and I connected. He adopted the motto "If you'll buy it, I'll grow it!" We developed a close relationship and became fast friends. I still source a good variety of beautiful produce from his farm, but Shawn and Wadmalaw Island have become my source for much more.

In 2006, I took an inspiring trip to California and ate a revelatory meal at Manresa. That meal changed my outlook on vegetables. Chef David Kinch, who gave me a tour of his garden, modeled the type of relationship between a cook and the soil that I wanted to emulate. I had a gut feeling that to evolve as a chef, I had to understand the dirt. So I started with a small patch in my backyard and I began lobbying McCrady's owners to lease a plot for me to experiment on. They told me to do some research,

find some land, and put a proposal together, which naturally led me back to Shawn.

Shawn is the kind of guy that you can cut a deal with over a couple of cold beers and a handshake, and on Wadmalaw such deals are usually made at the Supper Club. From the outside, Wadmalaw Supper Club, as it's formally called—if anything about the club besides the required recitation of the Pledge of Allegiance before dinner can be called "formal"—doesn't seem like anything particularly special. It's a bunch of farmers and good ol' boys, natives of Wadmalaw and all-around proud Americans, eating tomatoes fresh from the field with tin cans of sardines in mustard and saltine crackers on a Wednesday evening. But in the context of farming, it's more. It's where Shawn first took me to meet his friends, the people he calls when a tractor is stuck three feet deep in the mud or a few extra hands are needed to put up a new pole shed. It's also a place to share ideas. Although you'd think that fellows like this would be wary of an outsider, they took me in as one of their own. Over cold beers, next to a life-size cardboard cutout of Dale Earnhardt Sr. and a makeshift kitchen with a rickety stove, we reinforce our ties.

This is how I've learned to understand the Lowcountry. For a couple of years, I leased a small piece of Wadmalaw from Shawn and he taught me how to farm—the good and the bad, the success and the failures. He also gave me a real appreciation of what it takes to cultivate good food, and, perhaps, the good life as well. For that, I owe him forever.

STRAWBERRY GAZPACHO

with TOMATO WATER JELLY, BASIL ICE, and STONE CRAB SALAD

SERVES 8

Unusual combinations of ingredients can create great interest in a dish. This one begins with the notion of traditional cold gazpacho, but I substitute sweet strawberries for the savory tomatoes. It's a great example of adapting a recipe to the best of what you find in your local market. Use your imagination; rather than strawberries, use cucumbers or fennel or any other ingredient that would pair nicely with the herbal ice and the rich stone crab.

Note that the tomato water must drain overnight and the basil ice has to be frozen overnight.

TOMATO WATER JELLY

10 sheets silver-strength gelatin (see Resources, page 326)

½ cup water

5 very ripe large heirloom tomatoes (about 3 pounds), cored and cut into chunks

Kosher salt

BASIL ICE

1 pound basil

3 tablespoons agave nectar

1 cup ice water

GAZPACHO

6 cups chopped strawberries (about 2 pounds)

1 cup diced red onion

1 cup diced red bell pepper

1 cup diced peeled English cucumber

1 garlic clove, minced

1 tablespoon chopped tarragon

¼ cup raspberry vinegar

¾ cup extra-virgin olive oil

CRAB SALAD

1 pound fresh stone crab meat, gently picked free of any shells

Juice of 1 lime

1 tablespoon extra-virgin olive oil

Kosher salt

Cayenne pepper

Sorrel sprigs

Basil, chervil, and tarragon leaves

Fennel fronds and flowers

EQUIPMENT

Butcher's twine

Two 9-by-12-inch baking pans

FOR THE TOMATO WATER: Place 3 of the gelatin sheets and 2½ cups of cold water in a bowl and let the gelatin soften, 5 to 10 minutes.

Warm the ½ cup water in a medium saucepan. When the gelatin sheets are soft, lift them from the water and gently wring them out. Add them to the warm water and heat over medium-low heat, stirring, until the gelatin has dissolved, about 1 minute. Remove the saucepan from the heat.

Puree the tomatoes in a food processor. Season with salt. Add the dissolved gelatin and pulse a few times to combine. Pour the tomato puree into ice cube trays and freeze until solid.

Set a strainer over a deep bowl and line it with a double layer of cheesecloth. Put the frozen tomato cubes in the strainer, gather up the edges of the cheesecloth, and tie with butcher's twine to form a sack. Attach the sack to a dowel or the handle of a wooden spoon, remove the strainer, and suspend the sack over the bowl. Refrigerate overnight. Clear liquid (tomato water) will drip into the bowl for about 24 hours; if any red liquid begins to appear, remove the sack (do not squeeze the cheesecloth, or the tomato water will cloud).

MEANWHILE, FOR THE BASIL ICE: Pick the basil leaves from the stems; discard the stems. Bring a large saucepan of salted water to a boil. Make an ice bath in a large bowl with equal parts ice and water. Put the leaves in a strainer and submerge them in the boiling water for 7 seconds. Remove and submerge them in the ice bath until completely cold. Drain, shake off the excess water, and dry on paper towels.

Put the basil in a blender and add the agave nectar. Turn the blender on, add the 1 cup ice water, and blend on high until smooth, about 7 minutes. Pour into a 9-by-12-inch baking pan. Freeze overnight.

MEANWHILE, FOR THE GAZPACHO: Working in batches if necessary, combine all of the ingredients except the olive oil in a blender and blend on high until very smooth. With the blender running, slowly drizzle in the olive oil. Transfer to a container, cover, and refrigerate until chilled. *(The gazpacho can be refrigerated for up to 2 days.)*

FOR THE TOMATO WATER JELLY: Put the remaining 7 gelatin sheets and 2 cups cold water in a bowl and let the gelatin soften, 5 to 10 minutes.

Warm ½ cup of the tomato water in a medium saucepan. When the gelatin sheets are soft, lift them from the water and gently wring them out. Add them to the warm tomato water and heat over medium-low heat, stirring, until the gelatin has dissolved, about 1 minute. Remove the saucepan from

the heat, add the remaining tomato water, and stir until well combined.

Line a 9-by-12-inch baking pan with plastic wrap, leaving an overhang on the two long sides. Pour in the tomato water, and refrigerate until gelled, about 2 hours.

Scrape the basil ice with a fork until it is slushy. Cover tightly and freeze. *(Tightly covered, the basil ice will keep for up to 1 week in the freezer.)*

JUST BEFORE SERVING, MAKE THE CRAB SALAD: Gently mix the crabmeat, lime juice, and olive oil in a bowl. Season with salt and cayenne pepper. Cover and refrigerate.

TO COMPLETE: While it is still in the pan, use a fork to rake the tomato water jelly into pieces. Divide the crab and diced jelly among eight chilled soup bowls, arranging the jelly next to the crab. Pour the soup around the crab. Garnish each bowl with sorrel, herbs, and fennel fronds and flowers. Scrape about 1 tablespoon of the basil ice into each bowl and serve immediately.

BEET AND STRAWBERRY SALAD WITH SORREL AND RHUBARB VINAIGRETTE

SERVES 6

I love the way the colors and flavors of the different earthy, salt-baked beets play off each other in this salad. The strawberries add sweetness and balance. Sorrel adds a nice acidic punch, while the rhubarb vinaigrette adds its own mouth-puckering sourness. Feel free to substitute any perfectly fresh green for the sorrel; I like to use peppery arugula when sorrel isn't available.

3 cups kosher salt

1 large egg white, lightly beaten

18 baby beets, various colors and varieties, scrubbed and tops trimmed to 1 inch

6 large stalks rhubarb, peeled and diced

5 dashes rhubarb bitters (see Resources, page 326)

1 tablespoon raspberry vinegar

8 strawberries, hulled and quartered

Extra-virgin olive oil

Maldon or other flaky sea salt

Freshly cracked black pepper

18 sorrel leaves, preferably a mix of wood sorrel, sheep sorrel, green sorrel, and/or red sorrel, washed and dried

EQUIPMENT

Juice extractor

Preheat the oven to 400°F.

Mix the kosher salt with the egg white in a bowl and spread in an even layer on a rimmed baking sheet. Put the beets on the salt mixture and turn to cover them completely with salt. Roast the beets for about 1 hour, until fork-tender. Remove from the oven.

When the beets are cool enough to handle, remove the skin and tips by rubbing them with a kitchen towel. Wipe off any salt that remains on the beets. Cut in half and reserve.

Run the rhubarb through a juice extractor and pour the juice into a stainless steel saucepan. Bring to a boil over medium-high heat, and cook until the juice is reduced to ½ cup, about 10 minutes. Add the bitters and vinegar, scraping the bottom of the saucepan to loosen and incorporate any thickened juices. Strain, cool to room temperature, and refrigerate until chilled.

TO COMPLETE: Place 6 beet halves on each of six plates. Place 5 strawberry quarters on each. Add the sorrel leaves. Drizzle the plates with the rhubarb juice and drizzle olive oil into the rhubarb juice. Sprinkle with Maldon salt and cracked pepper and finish each plate with 3 sorrel leaves.

SALAD OF PLUMS AND TOMATOES WITH RASPBERRY VINEGAR, GOAT CHEESE, AND ARUGULA PESTO

SERVES 6

A few years ago, I bought some plums and tomatoes from a farmer at the market. When we loaded them in my truck, I was struck by how great they smelled together. Back in the kitchen, the first thing I did was dress a few plums and tomatoes with a bit of raspberry vinegar—what a revelation! A great raspberry vinegar that has a very low acidity brings out the fruitiness of the ripe tomatoes and marries well with the slight acidity of the plums. The vegetal bite and texture of the arugula pesto brings the sweet bits together very nicely.

ARUGULA PESTO

¼ cup pine nuts

1 pound arugula, washed and patted dry

2 ounces Pecorino Romano cheese, grated

1 garlic clove, minced

1 teaspoon sugar

½ cup grapeseed oil

¼ cup extra-virgin olive oil

Kosher salt and freshly ground black pepper

4 large heirloom tomatoes (about 6 pounds)

3 ripe plums, cut away from the stones and thinly sliced

½ cup raspberry vinegar

½ cup extra-virgin olive oil

Kosher salt and freshly ground black pepper

8 ounces goat cheese, crumbled

3 ounces baby arugula, washed, patted dry, and stems removed

Arugula blossoms (optional)

FOR THE ARUGULA PESTO: Lightly toast the pine nuts in a skillet over medium heat, about 2 minutes. Watch carefully and stir occasionally to keep them from burning. Transfer to a plate and cool.

Put the pine nuts, arugula, cheese, garlic, and sugar in a blender and blend on high until smooth, about 2 minutes. With the blender running, slowly pour in the oils. When the oils are incorporated, season with salt and pepper. Set aside at room temperature. *(The pesto can be made up to 2 days ahead. Transfer to a small container, pour a very thin layer of olive oil over the top to prevent the arugula from discoloring, cover tightly, and refrigerate; bring to room temperature before using.)*

Cut three of the tomatoes into interesting shapes: slices, wedges, and rectangles. Lay the tomatoes and plums on a rimmed baking sheet in a single layer. Drizzle with the raspberry vinegar and olive oil and season with salt and pepper.

TO COMPLETE: Smear ¼ cup of the pesto in a circle on each of six plates. Place slices, wedges, rectangles, and cubes of tomato on each plate. Arrange several plum slices over the tomatoes on each plate. Sprinkle the goat cheese over the plates and garnish with the baby arugula and, if you have them, arugula blossoms.

NOTE

This recipe makes more pesto than you'll need. Freeze the leftover in a container, with a layer of olive oil as above, for later use.

WATERMELON AND RED ONION SALAD WITH BIBB LETTUCE, PICKLED SHRIMP, AND JALAPEÑO VINAIGRETTE

SERVES 6
AS AN APPETIZER
OR A LIGHT LUNCH

Pickled shrimp are an old standby in Charleston; you'll see them all around town at dinner parties and as a bar snack. They date back to at least the late eighteenth century, when Harriot Pinckney Horry included them in her plantation receipt book. You can serve the shrimp on their own—we pass them as bar snacks at Husk—or pair them with preparations like this watermelon salad, making a dish substantial enough for a light lunch. I always get a few strange looks when I pair watermelon with savory ingredients, but the acidity of something like pickled shrimp goes really well with the sweetness of the fruit.

Note that the pickled shrimp must cure for at least 2 days.

PICKLED SHRIMP

2 cups white vinegar
1 cup fresh lemon juice
½ cup fresh lime juice
¼ cup fresh orange juice
¼ cup extra-virgin olive oil
1 garlic clove, minced
1 tablespoon kosher salt
1 tablespoon coriander seeds
1 tablespoon yellow mustard seeds, crushed
1 teaspoon celery seeds
1 teaspoon fennel pollen
½ teaspoon crushed red pepper flakes
½ teaspoon turmeric
1 fresh bay leaf
1½ pounds large shrimp (16–20 count), peeled and deveined

JALAPEÑO VINAIGRETTE

3 jalapeño peppers, seeded and diced
Grated zest of 1 lime (use a Microplane)
½ cup fresh lime juice
1½ teaspoons sugar
1½ teaspoons kosher salt
¾ cup canola oil
¼ cup extra-virgin olive oil

1 head Bibb lettuce, separated into leaves, washed, and patted dry
1 small ripe watermelon, peeled, seeded, and cut into 1-inch cubes
1 small red onion, shaved paper-thin and held in ice water

FOR THE SHRIMP: Put all of the ingredients except the shrimp in a medium stainless steel or enameled pot, bring to a simmer over medium-high heat, and cook for 10 minutes. Remove from the heat and add the shrimp. Allow the shrimp to poach in the pickling liquid, uncovered, until they turn pink and begin to curl, about 15 minutes.

Remove the shrimp and refrigerate. Pour the pickling liquid into a sterilized glass or stainless steel container and let cool to room temperature.

Return the shrimp to the pickling liquid, cover, and refrigerate for at least 2 days to cure. *(Tightly covered, the shrimp will keep for up to 2 weeks in the refrigerator.)*

FOR THE VINAIGRETTE: Put the peppers, lime zest, lime juice, sugar, and salt in a blender and blend until smooth, about 5 minutes. With the blender running, slowly add the oils and blend to emulsify.

TO COMPLETE: Put the lettuce in a large bowl and gently toss it with enough vinaigrette to lightly coat the leaves. Divide the lettuce, shrimp, and watermelon cubes among six salad bowls or plates. Garnish each with the sliced red onion.

=== **NOTE** ===

The recipe makes more vinaigrette than you will need for this dish, but it will keep, tightly covered, for up to 5 days in the refrigerator. Whisk, or shake if stored in a jar, before using.

SALAD OF PEA SHOOTS AND WATERCRESS WITH CURED EGG YOLK, RADISH, AND BUTTERMILK DRESSING

SERVES 6

After a long winter of cooking root vegetables and leafy greens, I am always eager for the first signs of spring. Look for pea shoots at a farmers' market—or convince a local specialty grower to snip a few for you from their peas in the early spring. Then bring him or her this simple delicious salad; I'd bet you'll never have a problem getting pea shoots again.

DRESSING

½ cup sour cream

½ cup whole-milk buttermilk

Generous 1½ ounces Pecorino Romano cheese, grated

¼ cup mayonnaise, preferably Duke's (see Resources, page 326)

1 tablespoon cider vinegar

1½ teaspoons Worcestershire sauce

½ cup chopped basil

1½ teaspoons chopped flat-leaf parsley

1½ teaspoons chopped chives

1½ teaspoons sugar

1 small garlic clove, minced

1½ teaspoons kosher salt

1½ teaspoons freshly ground black pepper

4 radishes with greens attached (any variety will do)

1 pound pea shoots, washed and dried

1 pound watercress, preferably wild, washed and dried

Kosher salt and freshly ground black pepper

2 Cured Egg Yolks (page 234)

FOR THE DRESSING: Combine the sour cream, buttermilk, cheese, mayonnaise, vinegar, Worcestershire, basil, parsley, chives, sugar, garlic, salt, and pepper in a bowl and mix well. Cover tightly and refrigerate for at least 4 hours, or preferably overnight, to allow the flavors to develop. *(Tightly covered, the dressing will keep for up to 5 days in the refrigerator.)*

TO COMPLETE: Remove the greens from the radishes, wash the greens, and dry them on paper towels. Using a mandoline or a very sharp knife, shave the radishes very thin.

Combine the radish greens, pea shoots, and watercress in a large bowl. Add ½ cup of the buttermilk dressing and toss the greens gently until well coated. Season with salt and pepper.

Arrange the salad in the center of six plates. Garnish with the shaved radishes. Using a Microplane, grate the cured egg yolks over the top.

=== NOTE ===

Leftover buttermilk dressing makes a great dip for just about everything from vegetable crudités to fried pickles.

GARDEN LETTUCES AND GREEN STRAWBERRIES

WITH MUSTARD DRESSING, SWEET ONIONS, AND FRIED PORK RILLETTES

SERVES 6

My friend New Orleans chef Donald Link once made me a salad of greens garnished with rabbit rillettes, which inspired this salad that we serve at Husk. Our take includes green strawberries, which bring a tartness and firm texture to the plate, and it's become a customer favorite. Most people know about the sweet onions that are grown in Vidalia, Georgia, but in Charleston, we use our favorites from Wadmalaw Island, and they're every bit as good.

PORK RILLETTES

1 pound boneless pork shoulder

1½ teaspoons kosher salt

1½ teaspoons sugar

1½ teaspoons freshly ground black pepper

2 cups Rendered Fresh Lard (page 316), melted

2 thyme sprigs

1 fresh bay leaf

6 cups canola oil, for deep-frying

1½ cups all-purpose flour

1 cup whole-milk buttermilk

1½ cups panko bread crumbs, finely ground in a food processor

Kosher salt

Cayenne pepper

MUSTARD DRESSING
(MAKES A GENEROUS 1 CUP)

½ cup mayonnaise, preferably Duke's (see Resources, page 326)

¼ cup whole-grain mustard

¼ cup Dijon mustard

1½ tablespoons cider vinegar

1½ teaspoons fresh lemon juice

½ teaspoon Worcestershire sauce

½ teaspoon Husk Hot Sauce (page 238) or Crystal hot sauce (see Resources, page 326)

½ teaspoon celery seeds

Kosher salt and freshly ground black pepper

SALAD

1½ pounds mixed lettuces

Kosher salt and freshly ground black pepper

20 green strawberries, washed, hulled, and sliced lengthwise ¼ inch thick

1 large sweet onion, shaved paper-thin and held in ice water

FOR THE PORK RILLETTES: Preheat the oven to 200°F.

Season the pork shoulder with the salt, sugar, and pepper and place it in a small pot that just holds it comfortably. Cover it with the pork fat and top with the thyme and bay leaf. Cover the pot and slow-roast the shoulder for 8 hours. Cool to room temperature.

MEANWHILE, FOR THE MUSTARD DRESSING: Whisk the mayonnaise, both mustards, the vinegar, lemon juice, Worcestershire sauce, hot sauce, and celery seeds together in a small bowl. Season with salt and pepper, cover, and refrigerate. *(Tightly covered, the dressing will keep for up to 5 days in the refrigerator.)*

Remove the cooled pork from the fat. Strain and reserve the fat. Shred the meat and put it in the bowl of a stand mixer fitted with the paddle attachment. Start the mixer on the lowest speed, gradually increasing it to medium-low speed. With the mixer running, slowly drizzle in 1½ cups of the reserved fat (reserve the remaining fat for another use). Check the seasoning. Refrigerate the pork until thoroughly chilled, at least 2 hours.

Roll the cold pork mixture into 1-inch balls. Place them on a baking sheet and freeze until firm, about 1 hour.

WHEN READY TO FRY THE RILLETTES, pour the canola oil into a deep fryer or a large pot and heat it to 350°F. Meanwhile, make a breading station by setting up three separate shallow bowls for the flour, buttermilk, and panko; season them all with salt and cayenne.

Working in batches, coat the rillettes in the flour, gently shaking off the excess, dip them into the buttermilk, allowing any excess to fall back into the bowl, and coat them in the panko, gently shaking off any excess; put them on a plate. You will need 24 rillettes for the salad.

Fry the rillettes, in batches, until golden brown and hot through, about 5 minutes. Check the interior with a cake tester or the tip of a knife to be sure they are hot. Drain them briefly on a rack covered with paper towels.

TO COMPLETE: Toss the lettuces with enough of the dressing to lightly coat them. Season with salt and black pepper. Arrange the lettuces in the center of six plates. Garnish the salads with the strawberries and onions. Place 4 rillettes around each salad.

BUTTER-BRAISED ASPARAGUS WITH STONE CRAB, NASTURTIUM CAPERS, AND CANE VINEGAR CREAM

SERVES 6

The Lowcountry asparagus season is fleeting, usually from Easter to Mother's Day. And the asparagus crop comes in about the same time we get our first stone crab claws. Every spring when these stars align, we pair the super-fresh asparagus with our amazing stone crab in as many ways as possible. I added the pickled nasturtium buds after reading about them in a nineteenth-century book called *The Complete Confectioner*. The cookbook says they can be "used instead of capers," which can be substituted here with good results.

1 cup heavy cream

1 tablespoon cane vinegar

Kosher salt

3 tablespoons unsalted butter

30 medium asparagus spears, tough ends cut off and bottom thirds of spears peeled

8 ounces fresh stone crab meat, gently picked free of any shells

30 Nasturtium Capers (page 224) or capers

6 organic nasturtium flowers, plus 18 small nasturtium leaves

Whisk the cream in a bowl until it starts to thicken. At this point, it is important to slow down and pay close attention to the thickness of the cream as you whip it: you can easily overwhip cream, and it will separate. When the cream forms stiff peaks, stop whisking. Use a rubber spatula to fold in the vinegar and a pinch of salt. Reserve the cream on the countertop.

Add the butter to a skillet large enough to hold all of the asparagus in one layer, set over medium heat, and heat until the butter melts and becomes foamy, about 2 minutes. Add the asparagus to the skillet in a single layer, season with salt, and shake the skillet to coat the asparagus with the butter. The asparagus will cook rather quickly, so don't leave the stove; shake the skillet every once in a while to turn the asparagus so it cooks evenly. This should take about 5 minutes. Transfer the asparagus to a plate lined with paper towels and keep warm on the stove.

Off the heat, add the crabmeat to the skillet and stir gently to coat it with the butter remaining in the pan and just warm it through; you should not have to put the pan back over the heat.

TO COMPLETE: Divide the asparagus among six plates. Top with 5 nasturtium capers. Garnish with the nasturtium flowers and leaves. Place a dollop of the cane vinegar cream off to the side and serve immediately.

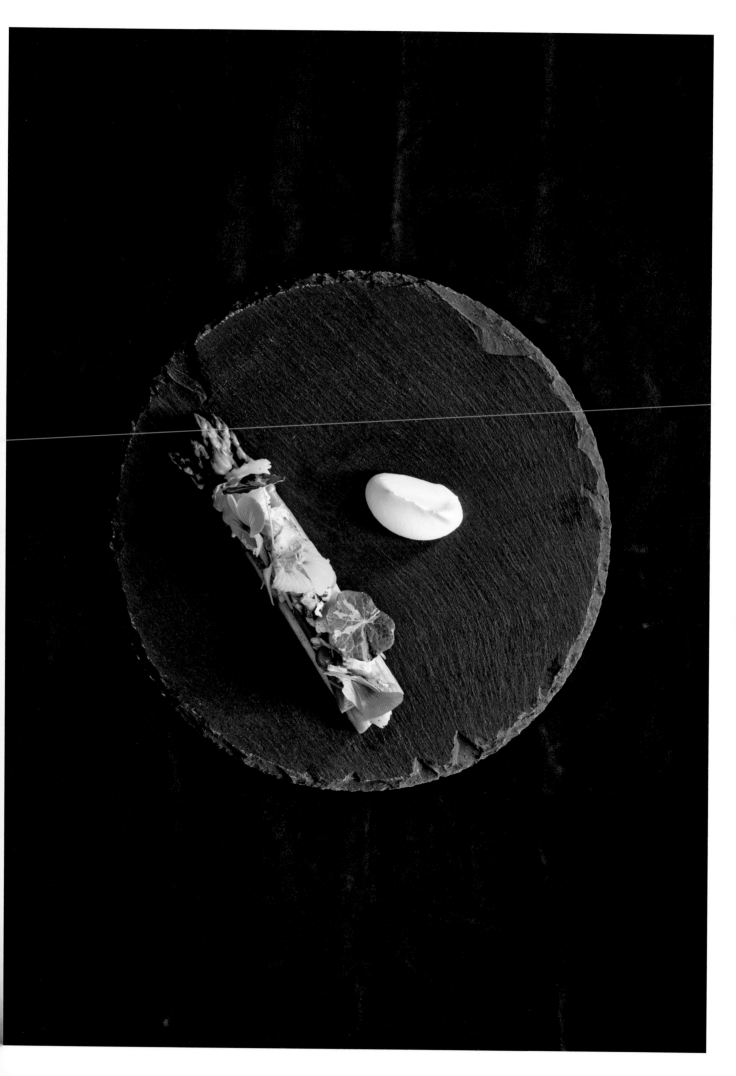

CARROTS BRAISED AND GLAZED IN CARROT JUICE

SERVES 6

We developed this technique of braising vegetables in their own juices to intensify the inherent flavors of the vegetables, similar to the way French chefs develop rich sauces by using bones and trimmings to make a dense broth. By cooking with the vegetables' own juices and reducing them down to a thin glaze, you can achieve incredible flavors. This technique works with all sorts of vegetables and even fresh fruits. Think beets poached in beet juice, strawberries poached and glazed in strawberry juice . . . you get the idea.

42 baby carrots with tops
Juice of 1 orange
2 tablespoons unsalted butter
1 tablespoon chopped tarragon
1 tablespoon chopped chervil

EQUIPMENT
Juice extractor

Remove the tops from the carrots. Pick the 24 nicest-looking tops and reserve in ice water to use as garnish.

Peel and rinse the carrots. Run 12 of the carrots through a juice extractor. Remove the pulp and run it through the extractor 3 or 4 more times to extract all the juice.

Place the remaining carrots, the orange juice, and carrot juice in a large saucepan set over medium-high heat, cover, bring to a simmer, and cook until the carrots are fork-tender, 6 to 8 minutes. Increase the heat to high to begin reducing the liquid. When the liquid is reduced to a glaze, after about 3 minutes, remove from the stove, add the butter, and stir to emulsify it with the reduced juices. Add the chopped herbs and stir to coat the carrots.

TO COMPLETE: Divide the carrots among six warm plates. Drizzle with the remaining glaze and garnish each plate with 4 carrot tops.

HEIRLOOM POTATO CONFIT

SERVES 12 TO 15

One year at potato-planting time, we placed an order with Celeste Albers to ensure that both restaurants stayed stocked throughout the season. When she finally delivered the potatoes, they came in at hundreds of pounds, so we used this technique to preserve them. Cooking them and storing them in the fat makes them even better! These potatoes can be used in a number of ways. My favorite method is a simple one: I cut them in half and roast them on the stovetop in their own cooking fat, rolling them around so they get nice and brown. You can finish them with a bunch of fresh herbs sprinkled into the pan just before serving.

BRINE
1 gallon water
2 cups kosher salt
¾ cup sugar

5 pounds small heirloom potatoes, washed
1 pound unsalted butter
2 cups extra-virgin olive oil
2 cups Rendered Fresh Lard (page 316)
1 cup rendered bacon fat
2 tablespoons kosher salt
1 teaspoon freshly ground white pepper
20 thyme sprigs
12 garlic cloves
2 fresh bay leaves

FOR THE BRINE: Combine 4 cups of the water, the salt, and sugar in a large stainless steel or enameled pot and bring to a simmer, stirring to dissolve the salt and sugar. Add the remaining 3 quarts water and stir. Add the potatoes, remove the pot from the stove, and brine the potatoes at room temperature for at least 6 hours, or overnight.

Preheat the oven to 250°F.

Combine the butter, olive oil, lard, and bacon fat in a Dutch oven and heat over medium heat until melted. Stir, add the salt, white pepper, thyme sprigs, garlic cloves, and bay leaves, and heat for 7 minutes to infuse the fat with flavor.

Meanwhile, remove the potatoes from the brine (discard the brine) and pat dry with a kitchen towel. Carefully place the potatoes in the hot fat, cover the Dutch oven, transfer to the oven, and roast the potatoes for 3 hours, until very soft. Cool the potatoes to room temperature.

The potatoes can be eaten right away, but they are better if they are refrigerated, covered with the cooking fat, in an airtight container for at least 3 days.

=== NOTE ===

Covered in the fat, the potatoes will keep for up to 1 month in the refrigerator. The longer they sit, the better they get. Because of bacteria on your hands, always use a spoon to retrieve the potatoes from the fat; make sure to cover the remaining potatoes with the fat.

EGGPLANT BARIGOULE

SERVES 6

As a nod to the many French Huguenots who fled religious persecution in France and helped to populate Charleston in the late 1600s, we cook this Southern take on a Provençal dish. *Barigoule,* traditionally made with artichokes, is a great technique that can be used for just about any vegetable imaginable, and I've even used the recipe as a base to braise chicken or beef. This eggplant version can be served over rice as a main course or as a perfect summer meal with a lightly grilled piece of fish on top. The leftover braising liquid also makes a great sauce.

Cynar is an Italian aperitif made from artichokes, but you can substitute dry vermouth if necessary.

2 tablespoons canola oil, plus more if needed

6 cups peeled, seeded, and diced (½-inch) eggplant (1 large eggplant)

Kosher salt

1 tablespoon extra-virgin olive oil

¼ cup diced country ham, such as Benton's (see Resources, page 326)

1 small shallot, cut into ⅛-inch dice

1 garlic clove, sliced paper-thin

½ cup peeled, diced (⅛-inch) carrots

½ cup peeled, diced (⅛-inch) celery

½ cup dry white wine

¼ cup Cynar or dry vermouth

1 cup Chicken Stock (page 318)

Juice of ½ lemon

Freshly ground black pepper

Set a large deep skillet over high heat and add the canola oil. When the oil begins to smoke, add a single layer of eggplant (you will have to work in batches). Season with salt and sear the eggplant on one side, without moving it, until it's golden brown, about 3 minutes. Turn the pieces over and brown on the other side, another 3 minutes. Transfer the eggplant to a plate. Repeat with the remaining eggplant. It is important to keep the temperature very high, so you may have to replace the canola oil between batches if it becomes too dark.

Reduce the heat to medium, add the olive oil, country ham, shallots, and garlic to the skillet, and cook, stirring occasionally, until the shallots and garlic soften, about 4 minutes. Add the carrots and celery, increase the heat to high, and cook until the carrots are lightly softened, about 2 minutes. It's important to not overcook the carrots: you want them to still have some texture when the braise is complete.

Return the eggplant to the skillet, reduce the heat to medium, and cook for 2 minutes. Increase the heat to high, add the white wine and Cynar, and cook, stirring occasionally, until the liquid is reduced by half, about 6 minutes.

Add the chicken stock and bring to a simmer, then reduce the heat to low, cover, and cook for about 20 minutes, until the eggplant is nice and soft but still has some bite. Add the lemon juice, season with salt and pepper, and serve.

NOTE

This dish can be cooked up to 2 hours ahead and held at room temperature; before serving, reheat over medium-low heat until hot through.

CREAMED CORN

My grandmother made creamed corn the old-fashioned way: strip the kernels from the cobs, scrape all the milk from the cobs using an old box grater, add a little salt, and then process in Mason jars in a canner. These preserves would be saved for special occasions, like Thanksgiving dinner. At Husk, I gussy up the recipe a little with a bit more cream and butter. You can also serve this as a soup by adding a little milk to thin it out. Either fresh or preserved under glass, nothing says summer like sweet corn from the garden, even when you're eating it in the dead of winter.

8 ears corn, husked

1½ tablespoons extra-virgin olive oil

1 small shallot, thinly sliced

3 garlic cloves, thinly sliced

2 cups heavy cream

3 thyme sprigs, tied together with kitchen string

1 tablespoon unsalted butter

Kosher salt and freshly ground white pepper

Cut the kernels from the corn; set aside. Using a box grater, scrape the "milk" from the cobs into a wide bowl; set aside.

Heat the olive oil in a large skillet over medium heat. Add half of the corn kernels, the shallots, and garlic and cook, stirring frequently, until the shallots and garlic have softened considerably, about 7 minutes. Add the cream, bring to a simmer, and cook, stirring occasionally to prevent scorching, until thickened, about 15 minutes. Remove from the heat.

Working in batches if necessary, transfer the corn mixture to a blender and blend on high until completely smooth, about 5 minutes. Strain through a fine sieve into a saucepan.

Add the remaining corn kernels, the reserved "milk" from the cobs, the thyme, and butter to the pan, bring to a simmer over medium heat, and simmer until the creamed corn has thickened and the whole kernels are soft, about 10 minutes. Remove the thyme, season with salt and white pepper, and serve.

NOTE

The creamed corn can be made up to 2 hours ahead and held at room temperature; gently reheat over low heat. Leftovers will keep, covered, in the refrigerator for up to 3 days.

LEMONY MUSTARD GREENS WITH BENNE

SERVES 6

Here is an example of old and new coming together. This nineteenth-century recipe combines whole benne seeds with braised greens, and I added some Charleston Hots peppers to spice things up a bit (if you can't get them, cayenne is a good substitute). I recommend cooking all types of greens quickly, with just a simple sauté. It helps them retain their flavor rather than braising them into submission. Building the dish quickly in the pan allows the lemon and hot pepper to add a nice layer of complexity that offsets the pleasant bitterness of the benne.

1 large bunch mustard greens

1 tablespoon Anson Mills Antebellum Benne Seeds (see Resources, page 326)

2 tablespoons extra-virgin olive oil

2 garlic cloves, sliced paper-thin

Grated zest (use a Microplane) and juice of 1 lemon

¾ teaspoon kosher salt

½ teaspoon freshly ground white pepper

½ tablespoon dried Charleston Hots (peppers; see Resources, page 326)

Remove the stems and tough center ribs from the mustard greens. Make stacks of the leaves, roll them into cylinders, and cut them crosswise into very thin ribbons. Wash in a sink filled with cold water, changing the water several times, to remove any sand from the greens. Drain well and pat dry with kitchen towels.

Toast the benne seeds in a small heavy skillet over medium heat, watching carefully and stirring occasionally, until light brown and fragrant, about 5 minutes. Spread the seeds on a plate to cool.

Heat 1 tablespoon of the olive oil in a large skillet over medium heat until warm but not hot. Add the garlic and cook until lightly toasted, about 2 minutes. Add the mustard greens and toss to coat. Cook the greens until they are just wilted, about 8 minutes. Add the lemon zest and juice, salt, white pepper, and Charleston Hots and toss well.

Remove from the heat and add the remaining tablespoon of olive oil. Finish with a sprinkling of the benne seeds. Serve immediately.

FIELD PEAS AND BEANS

Beans have long been a staple crop of the Americas. Native Americans planted them alongside corn and squash as one of the "three sisters" that every kid learns about in school. Farmers use them to improve the soil and often plant "cowpeas" (so named because they are a favored fodder of grazing cattle) as a cover crop. Beans also fed the slaves, and many of the traditional bean dishes of the Lowcountry—like the hoppin' John of rice and field peas that brings us good luck on New Year's Day—can be attributed to the influence of those African Americans. I imagine that beans have stuck around for a few other reasons. For one, they taste good. I like to eat them raw with just warm oil and a bit of salt tossed on. But perhaps what really keeps them around is the fact that they keep so well. Dried field peas seemingly last forever, and the strips of goose beans or greasy beans hung up on a farmstead porch can replace meat in a time of scarcity.

Anyone in a warm climate with enough days and a few square feet of dirt can grow beans. In Charleston, you often see a few sticks of bamboo lashed together into a teepee of sorts strung vertically with baling twine and lavishly adorned with dark green leaves and fresh pods hanging from it, like an edible Christmas tree in the backyard.

You'll also see long rows of soybeans, used in rotation with corn. Most of these beans are destined for commercial oil production or animal feed, but some innovative growers are planting the Japanese edamame and other specialty heirloom beans sought out by chefs. That's really the wonder of the bean; it's tough to list all the kinds there are. They often define families; many clans have a distinctive variety passed down through generations of sharing.

One such variety of field peas is the one we call Sea Island red peas, and its history gives you a glimpse of the complexity of Lowcountry food. The Sea Island red pea reportedly came to America with Northern Italians from the Veneto, who were hired to build rice ditches before the arrival of African slaves skilled in the practice. The Africans then adopted the peas, and the Italian *risi e bisi* became what the Gullah people called *reezy peezy*. Who knows if this is truth or myth, but the story illustrates the importance of beans in Lowcountry culture. We can be sure that pots have been cooking down red pea gravy for as long as rice has been steaming over the fire on the banks of the Ashley River. The rice culture brought the wealth that built the city, and the peas that find their way into the pot are as integral as any other ingredient in Charleston's pantry. Beans are woven into the fabric of Lowcountry food, but if you look hard enough at your own city, whether Boston or Seattle, you'll likely find some beans in its culinary past as well.

POLE BEANS
WITH TOMATOES
AND BENNE

SERVES 6

Tomatoes, beans, and benne seeds are a classic combination in Lowcountry kitchens. The three flourish at the same time in the garden, and I live and cook by the idea that if the ingredients are in the same garden, they will be delicious on the same plate. A lot of great dishes are born that way, and this one is a timeless example. We refine the concept a bit, but feel free to get rustic and just serve the raw tomatoes over the beans, sprinkled with the benne. It'll be tasty, but cooking the tomatoes for eight hours really intensifies the flavor. Any leftover tomatoes can be used for a number of other dishes—a good trick to have up your sleeve.

3 heirloom tomatoes, cut crosswise in half, and seeded

3 tablespoons extra-virgin olive oil

Kosher salt and freshly cracked black pepper

2 tablespoons Anson Mills Antebellum Benne Seeds (see Resources, page 326)

3 pounds heirloom beans, such as greasy, cut-short, half runner, or Turkey Craw

4 cups Vegetable Stock (page 316)

Benne oil (optional)

Set the oven as low as it will go: the best temperature for this technique is 190° to 200°F. Place a wire rack on a rimmed baking sheet.

Toss the tomatoes with 2 tablespoons of the olive oil and season with salt and cracked pepper. Place the tomatoes cut side down on the rack on the baking sheet. Slow-roast for at least 4 hours, and up to 8 hours. They should be bright red and shriveled. Cool to room temperature.

Gently remove the tomato skins and discard. Put the tomatoes in a baking dish and sprinkle with a little salt and cracked pepper. Put the dish in the turned-off oven; its residual heat should keep them warm.

Toast the benne seeds in a small heavy skillet over medium heat, watching carefully and stirring occasionally, until light brown and fragrant, about 5 minutes. Spread out on a plate to cool.

Remove any strings from the beans and cut the beans into bite-sized pieces. Put them in a large saucepan, cover with the vegetable stock, and add a sprinkle of salt and the remaining 1 tablespoon olive oil. Bring the beans to a simmer, cover, and cook over low heat until tender, about 30 minutes. The beans should remain just covered with liquid while they cook; add a little water if needed. (The beans can be cooked up to 30 minutes ahead, held at room temperature, and reheated over medium-low until hot through.)

Meanwhile, when the beans have almost finished cooking, check to make sure that the tomatoes are still warm. If they have cooled, turn the oven on for a couple of minutes to warm them.

Divide the beans among six warm plates and top each pile of beans with half of a tomato. Sprinkle with the benne seeds and benne oil, if using, and finishing salt and serve.

BAKED SEA ISLAND RED PEAS

SERVES 8

It's no secret that I am completely infatuated with American barbeque. I'm even a member of a barbeque team, The Fatback Collective. I love everything about the culture of "Q," especially the side dishes.

This is a spin on classic baked beans using Charleston's native Sea Island red peas. They are incredible, but I'm also biased, because they were one of my first seed-saving projects at my Wadmalaw garden (they've since been listed in Slow Food's "Ark of Taste" program, which helps farmers, chefs, and consumers identify heritage foods that are at risk of extinction). The peas have an incredibly earthy flavor that pairs nicely with a cold beer and some smoky barbeque. At Husk, we cook them in the smoker underneath the pork shoulders, so that they catch all of those juicy drippings. If you have a smoker, I suggest you do the same!

1 cup diced bacon, preferably Benton's (see Resources, page 326)

1 large sweet onion, cut into medium dice

1 large Anaheim pepper, cut into medium dice

2 cups Pork Stock (page 319)

1 cup BBQ sauce, preferably a western North Carolina–style sauce

¾ cup Kentucky bourbon

5 Pickled Ramps (page 233), roughly chopped

½ cup pitted Medjool dates, roughly chopped

3 tablespoons Rendered Fresh Lard (page 316), melted

2 tablespoons Husk Hot Sauce (page 238) or Crystal hot sauce (see Resources, page 326)

1 tablespoon Worcestershire sauce

1 tablespoon sorghum (see Resources, page 326)

1 tablespoon kosher salt

1 teaspoon freshly ground white pepper

1 teaspoon smoked paprika (see Resources, page 326)

1 teaspoon dry mustard

2 fresh bay leaves

1 pound Anson Mills Sea Island Red Peas (see Resources, page 326), soaked in a pot of water in the refrigerator overnight

Preheat the oven to 250°F.

Put the bacon in a large Dutch oven and cook over medium heat, stirring frequently so that it doesn't burn, until the fat is rendered. Increase the heat to medium-high and cook the bacon until crispy, 4 to 5 minutes.

Add the onion and Anaheim pepper and cook, stirring frequently, until soft, about 5 minutes. Add the pork stock, BBQ sauce, bourbon, ramps, dates, lard, hot sauce, Worcestershire sauce, sorghum, salt, white pepper, paprika, dry mustard, and bay leaves and stir to combine well. Drain the peas, add to the pot, and stir well.

Transfer to the oven and bake for 3½ hours, stirring every 30 minutes or so to make sure the peas aren't sticking, until the peas are soft and the sauce has thickened. (The peas can be cooked up to 3 days ahead and refrigerated; reheat in a 200°F oven until hot through.)

Serve hot.

ROASTED CAULIFLOWER
with MEYER LEMON and BROWN BUTTER, WATERCRESS, and PINK PEPPERCORNS

SERVES 6

Vegetarian dishes such as this one can be substantial enough to serve as an entrée, though this also makes a good appetizer or side. Roasted cauliflower fits as well alongside a perfectly poached piece of late-summer fish as it does heavier mains of game or fowl. Goat butter adds a goat-cheese funkiness—and another layer of flavor.

CAULIFLOWER

1 large head cauliflower with a 1-inch stem (about 2½ pounds)

Kosher salt

3 tablespoons unsalted butter

½ cup canola oil

About 3 cups Vegetable Stock (page 316)

½ cup heavy cream

Freshly ground white pepper

BROWN BUTTER SAUCE

8 ounces goat butter (if not available, substitute an Irish butter or Plugrá)

Juice of ½ Meyer lemon

1 teaspoon turmeric

3 Pickled Ramps (page 233), chopped

1 tablespoon chopped flat-leaf parsley

1 bunch watercress, washed, patted dry, and tough stems removed, for garnish

Grated zest of 1 Meyer lemon (use a Microplane)

1 tablespoon pink peppercorns

FOR THE CAULIFLOWER: Preheat the oven to 375°F.

Remove the green leaves from the cauliflower but leave the stem intact. Press a small ring mold or round cookie cutter into the bottom of the stem so that the head of cauliflower will stand upright. Generously sprinkle the cauliflower with salt. Heat the butter and oil in a large, deep ovenproof skillet over medium heat and stand the cauliflower up in the skillet. Cook, occasionally spooning the hot butter and oil mixture over the cauliflower, for about 20 minutes, or until the outside of the cauliflower is golden brown.

Transfer the skillet to the oven and roast the cauliflower for about 10 minutes, until fork-tender. Remove the cauliflower from the oven and let stand for 10 minutes. Turn off the oven and open the oven door to cool it down.

MEANWHILE, FOR THE SAUCE: Heat the butter in a small skillet over medium-high heat, stirring, until it is golden brown and starts to smell slightly nutty, about 8 minutes. Remove the skillet from the heat, add the lemon juice, and stir to emulsify.

Return the skillet to the heat and reduce the heat to low. Add the turmeric, ramps, and parsley and cook for 1 minute to blend the flavors. Keep the sauce warm at the back of the stove; if necessary, reheat gently over low heat before serving.

Grease a rimmed baking sheet. Remove the cauliflower from the ring mold. Remove the stem from the cauliflower and peel away the outside of it, reserving the peel, then slice the stem into the thinnest circles possible; set aside. Cut the cauliflower crosswise into ½-inch-thick slices. Reserve the 6 center slices for serving. Chop the remaining slices and any scraps and reserve. Lay the cauliflower slices on the prepared baking sheet and place it in the still-warm oven (with the door closed) while you finish the dish.

Put the peelings from the stem and the reserved scraps in a small saucepan, add enough vegetable stock to cover, and bring to a simmer. Add the cream and stir to combine. Transfer the mixture to a blender and blend on high to a very smooth puree, about 7 minutes. Season with salt and white pepper.

TO COMPLETE: Make a pool of cauliflower puree in the center of each of six warm plates. Place a cauliflower slice on top of the puree on each plate and top with a small mound of watercress. Garnish with the slices of stem. Spoon the sauce on the cauliflower and sprinkle with the lemon zest and pink peppercorns.

ANY-VEGETABLE SALAD—BRAISED, COOKED, AND RAW, WITH BENNE-RICE BROTH

SERVES 6
AS AN APPETIZER

The idea of this dish is to use the freshest vegetables on hand—to capture the flavors of a day in the garden. We serve the vegetables cooked, raw, and pureed, all on the same plate. Simply go to the market and choose six of the best-looking vegetables. The ones that catch your eye will taste the best when cooked too.

Note that while this salad is relatively simple to make, you'll need to start 2 days ahead to puff the rice.

BENNE-RICE BROTH
Kosher salt

1 cup Anson Mills Carolina Gold Rice (see Resources, page 326)

2 cups canola oil

3 tablespoons Anson Mills Antebellum Benne Seeds (see Resources, page 326), lightly toasted

1 fresh bay leaf

1 tablespoon unsalted butter

VEGETABLES
12 ounces each of the 6 best-looking vegetables at the market

4 cups Vegetable Stock (page 316)

Edible flowers

Herb sprigs and leaves

EQUIPMENT
Dehydrator

FOR THE RICE: Pour 4 cups water into a medium saucepan, add 2 teaspoons salt, and bring to a boil over high heat. Add the rice, stir once, reduce the heat to low, and simmer for 8 minutes, or until tender. Transfer the rice and liquid to a container and cool to room temperature, then cover and refrigerate overnight.

The next day, drain the rice, reserving the broth. Return the broth to the container, cover, and refrigerate.

Dry the rice overnight in a dehydrator at 180°F, until it is very dry. You should be able to snap the grains in half.

TO PUFF THE RICE: Heat the canola oil in a medium saucepan over high heat to 400°F. Remove from the stove and add 2 tablespoons of the dehydrated rice. The rice will puff and rise to the top. Drain the oil through a fine-mesh strainer into another saucepan. Transfer the rice from the strainer to a paper-towel-lined baking sheet and spread it in a single layer. Sprinkle with salt. Bring the oil in the saucepan back to 400°F and repeat the process. It will take 4 or 5 times to puff all of the rice. The rice can be held at room temperature for up to 6 hours.

FOR THE BROTH: Pour the reserved rice broth into a saucepan, add the benne and bay leaf, bring to a simmer, and simmer over very low heat for 20 minutes. Strain the broth into another saucepan and set aside.

FOR THE VEGETABLES: Cut most of the vegetables into bite-sized pieces, reserving a couple of whole ones. Cook each vegetable in a separate saucepan, in one even layer. Add just enough stock to cover the vegetables, lay a circle of parchment paper on top, and simmer over medium heat. Cook until fork-tender. Transfer to a baking sheet. Combine the cooking liquid from all the vegetables in a medium saucepan.

Pick three of the cooked vegetables to serve as is and three to puree. Put the vegetables to serve as is into their individual saucepans and cover with cooking liquid. Separately put each vegetable to puree in a blender with enough stock to blend until smooth, then transfer to a saucepan. Cover to keep warm.

Set the saucepan of rice broth over medium heat. Add the butter, tilt the saucepan toward you, and, with an immersion blender, aerate the mixture until bubbles form. When you remove the blender and set the saucepan down, the bubbles will rise to the top.

Reheat the vegetables and vegetable purees over low heat just until warmed through.

TO COMPLETE: Shave a few slices from each of the reserved vegetables on a mandoline. Place purees on each plate. Place the bite-sized vegetables in the center of the plates. Garnish with the shaved vegetables, flowers, herb sprigs, and leaves. Spoon some bubbles on top. Sprinkle with puffed rice.

BENNE

"Benne is the missing link to Lowcountry cooking," Glenn Roberts, owner of Anson Mills, once told me. Ever since then, I've been chasing down the lost West African sesame of older times, and digging through ancient "receipt" books to decipher what Glenn meant. I learned that sesame seeds were not garnishes or afterthoughts in the history of Charleston's food, but were grown as subsistence crops, used as protein and fat substitutes when meat was scarce. They represent a side of Charleston cuisine that is almost forgotten. Just as tahini constitutes a significant flavor profile of the eastern Mediterranean, benne seasons the pot of the historical Lowcountry.

Benne is a Mende word referring to *Sesamum indicum,* but its seed bears little relation to the sesame seeds that come on your fast-food burger bun. The benne of Charleston, the kind that gave rise to pralines and the sweet wafers so identified with the city, came over with the slave trade from the rice coast of Africa. In the nineteenth century, benne was every bit as important as okra—of which it's a close relative—in the African-American diet. It was typically browned or toasted to intensify its earthy, nutty flavor, and then ground or mashed to release the volatile oils before it was used to thicken stews and sauces, or lightly heated and pressed to make oil. Then the leftover mash would find its way into pots of collard greens and kale, a rich oleo in a land too warm for cheap butter.

The plant has its drawbacks, however. It is inefficient to harvest, and the search for alternative oils during the period of agricultural industrialization resulted in a switch to sesame seeds more suitable to that end. In the process, the nuances of flavor that made benne a staple, rather than a sidekick, gave way to our modern understanding of it as a sort of garnish. If you want the real thing, you'll have to order benne from Anson Mills (see Resources, page 326), or come to Charleston and talk to someone in the know.

Fortunately, I have access to growers who can give us the entire plant. We use the razor-like leaves and beautiful white flowers for garnishing plates. Sometimes, taking a cue from the favorite Southern practice of deep-frying okra, we fry young seed pods whole in a shatteringly light tempura batter. No matter how it's cooked, heirloom benne deserves more attention from professional and home cooks alike, and it is well worth searching out.

THE MILL

GOOD GRAINS

Growing up in Virginia meant eating lots of corn. Cornbread, corn pone, and grits were important parts of our cooking. Nashville offers a similar perspective, but Charleston presents something different. The South Carolina Lowcountry is where rice was born in America, and the plantation agriculture that it brought along profoundly shaped the people and their way of life. It also shaped my cooking.

It took me some time in Charleston to fully appreciate the significance of rice, how it contributed to the city's wealth, created winners and losers, and wove itself into the fabric of Charleston's history. Its story ultimately led me to mentors like Glenn Roberts and David Shields, expert seedsmen and researchers interested in the history of Lowcountry agriculture. My understanding of the rice plantation pantry was the starting point of my journey and I'm still in awe of its depth, which I've only begun to explore.

Antebellum plantations were where grain was grown, be it rice, corn, or wheat. Today we take the bags of bleached and bromated white flour on every grocery store shelf for granted. Most people don't know the difference between what's on grocery-store shelves and grains straight from the field because they've never tasted fresh-milled grains. I'm convinced that once people taste flavorful heirloom grains, they won't want to go back to the often genetically engineered versions sold in big grocery stores.

This is what Husk in Charleston is all about. We translate the plantation pantry of antebellum Charleston for a modern palate, using all of those lost flavors that others in the Lowcountry are

reintroducing and growing for our local market. Husk
may be known for pork butter, spicy fried pig's ears,
and bourbon, but all of that flavor starts in a field of
rice, that becomes one of field peas next season, and
maybe some rows of corn the next, a rotation of crops
that some antebellum planters designed to cycle over
more than a dozen years.

What started as an experiment in finding old
flavors of grains that used to grow around Charleston
has completely changed my way of looking at
food, and if you start cooking with heirloom grains,
you will also see the rich diversity that's been lost
to industry. Recipes like Wheat Thins (page 86),
Cracklin' Cornbread (page 71), and Rice Griddle
Cakes (page 86) will certainly work with regular
store-bought grain, but if you want to get the full
experience of the Old South, seek out authentic
ingredients. See Resources, page 326, for some of my
favorite purveyors.

CORN GRIST

Corn is prepared in more forms than you can count. Whether it is mashed up into gruel, or ground into cornmeal for hushpuppies and cornbread, the aroma of a kitchen changes considerably when corn gets involved. It simply starts to smell Southern.

CORNMEAL

Cornmeal is ground dried corn. You can find it at any chain grocery store, but the best cornmeal is stone ground and you may need to source that at a mill or through mail order. Unlike more modern methods of milling, the stones don't substantially heat up the grains, resulting in a superior flavor and texture. Find a miller who grinds corn to order or purchase a small home grinder that will allow you to vary the size of the grind. Many millers ship grains via mail order, and a home mill can be easily purchased on the Internet.

HOMINY

If you take dried flint corn and cook it in lye until the outer hull of the kernel separates, you'll leave the germ of the kernel behind—and get hominy. This process, called nixtamalization, originated with Mesoamerican Indians and has a very specific effect: it unlocks the nutritional power of corn, making it much more digestible, especially when combined with rice or beans. Unlike Native Americans, Southerners and Europeans didn't fully adopt this practice and thus they often lacked the complete nutritional protein that it creates, leaving their populations who subsisted on cornmeal and preserved meat susceptible to a vitamin deficiency disease called pellagra.

HOMINY GRITS

Old-timers call it "little hominy," but modern commercial grits bear little resemblance to the staple grist of yesteryear. Industrial milling and commercial corn production mean that most of the grits you find are simply coarsely ground cornmeal, but hominy grits are nixtamalized dried kernels ground to a coarse consistency.

MASA

Latin Americans take fresh hominy and grind it while still wet, producing a soft corn flour that constitutes the basis for everything from tortillas to tamales. The commercial kind is called *masa harina* and comes dried in bags, like cornmeal or flour.

POLENTA

Corn traveled quickly to Italy after its "discovery" in the Americas, and it soon replaced buckwheat and farro as the grain of choice for polenta. Very similar to the South's grits and African *ugali,* polenta is a cornmeal mush originally eaten by peasants, a staple of the cuisine Italians call *la cucina povera,* but it is often made from flint corn, a very hard variety that has a lower starch content.

SAMP

Early colonists used the terms "grits" and "samp" interchangeably, but when we talk about samp today, we are referring to cracked hard flint corn. It's hard to make, since the best samp is cracked by hand, but the kernels of good samp can be shattered, producing very little corn flour in the process. This type of rough corn cooks up like rice, tender and fluffy.

CRACKLIN' CORNBREAD

MAKES ONE
9-INCH ROUND LOAF

My favorite ball cap, made by Billy Reid, has a patch on the front that reads "Make Cornbread, Not War." I'm drawn to it because cornbread is a sacred thing in the South, almost a way of life. But cornbread, like barbeque, can be the subject of great debate among Southerners. Flour or no flour? Sugar or no sugar? Is there an egg involved? All are legitimate questions.

When we opened Husk, I knew that we had to serve cornbread. I also knew that there is a lot of bad cornbread out there in the restaurant world, usually cooked before service and reheated, or held in a warming drawer. I won't touch that stuff because, yes, I am a cornbread snob. My cornbread has no flour and no sugar. It has the tang of good buttermilk and a little smoke from Allan Benton's smokehouse bacon. You've got to cook the cornbread just before you want to eat it, in a black skillet, with plenty of smoking-hot grease. That is the secret to a golden, crunchy exterior. Use very high heat, so hot that the batter screeches as it hits the pan. It's a deceptively simple process, but practice makes perfect, which may be why many Southerners make cornbread every single day.

4 ounces bacon, preferably Benton's (see Resources, page 326)

2 cups cornmeal, preferably Anson Mills Antebellum Coarse Yellow Cornmeal (see Resources, page 326)

1 teaspoon kosher salt

½ teaspoon baking soda

½ teaspoon baking powder

1½ cups whole-milk buttermilk

1 large egg, lightly beaten

Preheat the oven to 450°F. Put a 9-inch cast-iron skillet in the oven to preheat for at least 10 minutes.

Run the bacon through a meat grinder or very finely mince it. Put the bacon in a skillet large enough to hold it in one layer and cook over medium-low heat, stirring frequently so that it doesn't burn, until the fat is rendered and the bits of bacon are crispy, 4 to 5 minutes. Remove the bits of bacon to a paper towel to drain, reserving the fat. You need 5 tablespoons bacon fat for this recipe.

Combine the cornmeal, salt, baking soda, baking powder, and bits of bacon in a medium bowl. Reserve 1 tablespoon of the bacon fat and combine the remaining 4 tablespoons fat, the buttermilk, and egg in a small bowl. Stir the wet ingredients into the dry ingredients just to combine; do not overmix.

Move the skillet from the oven to the stove, placing it over high heat. Add the reserved tablespoon of bacon fat and swirl to coat the skillet. Pour in the batter, distributing it evenly. It should sizzle.

Bake the cornbread for about 20 minutes, until a toothpick inserted in the center comes out clean. Serve warm from the skillet.

NOTE

Use any leftovers to make Cornbread and Buttermilk Soup (page 76).

CORNMEAL HOECAKES

MAKES ABOUT
12 SMALL CAKES

Hoecakes reference a time when people didn't have proper vessels for cooking. So hoecakes were made using whatever tools were available. Often it was a straightened "chopping hoe" that could be held over the fire, providing a flat surface on which to bake. Of course, if you didn't have a pan to cook in, you probably didn't have chemical leavening of any type either. When heated until very hot on the embers of a fire, the scorching metal would cause the water in the batter to expand very rapidly, giving a bit of rise to the finished cakes.

Made with a little pork fat stirred into a simple cornmeal-and-water mixture, hoecakes are one of the most evocative of working-class foods. Eat them plain with some sorghum drizzled over the top. If you're feeling fancy, top them with a salad of smoked trout and a few watercress leaves—or go completely over the top: dollop them with a bit of fresh trout roe, and serve with a bowl of thick buttermilk for spooning on the side.

2 cups cornmeal, preferably Anson Mills Antebellum Fine Yellow Cornmeal (see Resources, page 326)

1 teaspoon baking soda

1¾ cups plus 2 tablespoons water

3 tablespoons Rendered Fresh Lard (page 316), plus more if needed

2 tablespoons unsalted butter, plus more if needed

Combine the cornmeal and baking soda in a small bowl.

Bring the water to a boil in a medium saucepan over high heat. Remove the pan from the stove and stir in the cornmeal. Stir in 1 tablespoon of the lard.

Heat the butter and the remaining 2 tablespoons lard on a griddle pan, preferably cast iron, over high heat. Cooking in batches if necessary, spoon the batter onto the griddle to make cakes about 1 inch in diameter. When you see craters in the tops of the cakes, after about 2 minutes, flip them and cook for about 2 minutes on the other side. They should be golden and crispy on both sides. Wipe out the pan with paper towels between batches if the butter scorches, and add fresh butter and lard. Serve the hoecakes immediately.

HEIRLOOM CORN

I've got corn in my blood, literally and figuratively. From the small bit of Native American Cherokee blood to my predilection for saving jar upon jar of old corn seeds for my collection, sometimes I think I was simply born to grow corn, and I aim to preserve one variety every year. I think that would be a respectable contribution to the world thirty years down the road. Corn is the most widely grown crop in the Americas, and it has a rich history reaching back to the time of the Mayans (who domesticated the plant from a native Mesoamerican grass), but few people understand how deeply corn production reaches into our lives and how rare the older heirloom varieties have become. In the wake of industrialization, Americans have become accustomed to hybridized ears of sweet corn designed to be cooked and eaten fresh, and there's nothing wrong with that; I love a good ear of corn in the summertime. But most corn is grown to be dried, maybe ground, and it ends up in everything from grits to ethanol fuel.

Heirloom corn is actually responsible for most of my favorite things, starting with bourbon. Of course, it's also responsible for grits and cornbread, and I can't imagine life without those two things on the table.

One of the reasons for corn's value is its ability (when dried) to keep through the winter and into the spring. Growing corn and squirreling it away in one form or another was commonplace in the Old South and allowed the early settlers to survive. I find it somewhat ironic that European immigrants (and now their descendants) became so dependent on this American plant. The first crop I ever planted for seed-saving purposes was an old Indian variety, Jimmy Red Corn, and now I have a collection of more than 100 different types that I hope to never stop growing and sharing with others. The corn varieties I'm after are honest, edible products and when I find some I like, I save every old seed I can.

One of the interesting things I've noticed while collecting seeds is how specifically regional the older corn selections are. Each variety tells a story of its respective region and the people who have lived there, and I love it when I find myself in a discussion with someone about their favorites. John Koykendall of Blackberry Farm can tell you the history of Webb Watson. My friend Glenn Roberts waxes on for hours about Carolina Gourd Seed or Henry Moore corn. Ask Celeste Albers what she prefers, and she'll say John Haulk, named for a well-known farmer from Wadmalaw Island, where she grew up. Ask Gra Moore his preferred corn, and he'll name ten, saying that it's like asking him to pick his favorite child.

We trade seeds like kids used to trade baseball cards, and for us they're every bit as valuable. I hope this book inspires a few others to become connoisseurs of corn, and its many stories, as well.

HOW TO COOK GRITS LIKE A SOUTHERNER

Grits have played an important part in the history of the South because corn was a crop that could be grown easily in the region and dried and stored for winter consumption. A bowl of grits can tell the story of a region, a family, or a time period. Most regions in the South have a favored variety of corn with historical reference of some sort. Some people like their grits very thick; others prefer them very fine and soupy. Some use milk; some use water to cook them. Some have them for breakfast only and would never eat them for dinner. But in some places, like in the Lowcountry, people have been known to consume grits at three meals a day.

When I was a kid, I didn't love grits the way that I love them now. Where I grew up, we didn't have a mill and the grits I ate came from a box. I liked them just fine, but I wasn't in love with them. These days it's a different story. I remember the first time that I tasted grits freshly milled from heirloom corn and slowly cooked on the stove with care. It was a revelation, one of those moments where you realize how much you have to learn. Few things make me happier than a bowl of grits. Anytime I travel out of the South I find myself craving grits; they are my go-to comfort food.

Everyone has their own way of cooking grits. To be honest, the way I cook grits changes almost every time I cook them because each bag of grits cooks differently. It all depends on a few variables: the variety of the corn, the freshness of the grits, and the coarseness of the grind.

The particular variety of corn used plays a role too. It's important to know what the starch and sugar contents are. Each variety of heirloom corn has a different starch content—dent corn has less starch than flint corn. Corn that has a high starch content will need more water in order to hydrate properly. Corn with a high sugar content will need less butter before it goes on the plate. I've also found that the length of time the corn spends drying in the field makes a difference in how fast the grits cook. The coarseness of the grind will also determine the cooking time and the end texture. The finer the texture, the faster the grits will cook.

Of course it's best if the corn is used right after it's milled. If you don't have that luxury, keep the grits in a Ziploc bag in the freezer, where they will keep for a month or so.

Here are a couple of rules to follow when cooking grits:

- Each variety of corn has a different aroma and a different flavor. I like to eat grits that come from around the South and compare tastes. Try different ones until you discover your favorite.

- Always soak grits for at least 6 hours, preferably overnight. Keep in mind that you are essentially hydrating the dried corn when you are cooking it, and you want to cook it as quickly as possible—the faster the grits cook, the more corn flavor they will have. Soaking grits starts the hydration process, so it will take less time to cook them.

- Right before you cook grits, skim off any chaff and hulls that have floated to the top of the soaking water; I use a fine-mesh skimmer. The chaff and hulls will never soften, so the grits will taste undercooked and you'll find yourself cooking them to death.

- Use the soaking water to cook the grits. I want to taste the corn when I eat grits; if you use milk to cook them, you mute their flavor.

- Stir, stir, stir! We have a rule in my kitchen: If you walk past grits cooking, give them a stir; make sure you scrape the bottom of the pot. And make sure you keep the sides of the pot clean. If some of the grits stick to the sides when you stir and you leave them there, they will never cook, but they will eventually find their way into the cooked grits.

CRISPY POKE-CORNMEAL FRITTERS

MAKES 18 FRITTERS

Pokeweed only grows wild in the South. You can find it growing on the side of the road, and you have to pick it young and tender. When I was a kid, my mom would pull the car over and have us pick wild pokeweed. Then she would make these fritters.

Many people won't eat pokeweed because the mature plant is known to be poisonous, but generations have enjoyed it in the springtime—as long as the pokeweed is picked before it shows a hint of purple coloration.

4 cups fresh pokeweed (about 8 ounces)

2 cups cornmeal, preferably Anson Mills Antebellum Fine Yellow Cornmeal (see Resources, page 326)

2 tablespoons kosher salt

½ teaspoon baking powder

¼ teaspoon baking soda

¼ teaspoon cayenne pepper

1 large egg

1½ cups whole-milk buttermilk

¾ cup peanut oil or Rendered Fresh Lard (page 316), melted

Spicy Pepper Jelly (page 214)

EQUIPMENT
¾-ounce scoop

Preheat the oven to 300°F. Line two rimmed baking sheets with parchment paper.

Wash the poke leaves in a sink filled with cold water, changing the water twice, to remove any sand. Drain and dry with kitchen towels. Remove the stems. Make stacks of the leaves, roll them into cylinders, and cut them into very thin ribbons.

Combine the cornmeal, salt, baking powder, baking soda, and cayenne pepper in a large bowl. In a small bowl, whisk the egg into the buttermilk. Stir the wet ingredients into the dry ingredients to combine. Stir in the poke.

Heat a large cast-iron skillet over medium heat. Add 2 tablespoons of the oil. When the oil shimmers, use a ¾-ounce scoop to add scoops of batter to the skillet. Do not overcrowd the skillet; cook about 3 fritters at a time. Cook the fritters until their tops begin to have bubbles like a pancake, about 1 minute. Flip them over and cook for another 2 minutes, or until a cake tester inserted in the middle comes out clean. Transfer the fritters to a wire rack. Add another 2 tablespoons of oil to the skillet and repeat the procedure. After you have cooked 9 fritters, pour out the oil, wipe out the skillet, and start with fresh oil.

Once all fritters have been cooked, place them on the prepared baking sheets and place them in the oven for 10 minutes. When they are done, they should be crispy and cooked through.

Serve immediately, with the pepper jelly.

CAROLINA RICE

The exact year that rice was introduced to Charleston is still a bit of a mystery, but by the tail end of the seventeenth century, it had certainly taken hold. An old legend credits a ship captain sailing in from Madagascar with a bag of the seed that became the basis for the famous Carolina Gold variety. But the basis for most South Carolina rice varieties was likely rice brought in from West Africa by enslaved Africans predating this story by many years; it's something that current genetic research validates. One thing is for sure: Africans knew rice, rice built Charleston, and without either, neither the city nor its cuisine would be what it is.

Some of the first Charlestonians emigrated from Barbados, where their fathers and grandfathers established massive sugar plantations, covering the land with cane fields and the slaves to work them. Our subtropical climate was a bit too cold for sugarcane, but these early settlers knew the wealth of the Atlantic trade world. They found a suitable crop in rice, developed the industry into the premier source, and Charleston became one of the richest places in America because of it.

The importance of the crop led to its incorporation in all aspects of the culture, and when you visit Charleston, you'll see vestiges of the rice trade everywhere. The furniture companies make fancy "rice beds" that sit in nearly every upscale bed-and-breakfast in town. Gullah ladies, who set up shop on street corners, sell woven sweetgrass "winnowing" baskets that were once used to separate the rice grain from its chaff. And no proper Charleston meal would ever come to the table without the ubiquitous bowl of rice. Rice became so important to Charleston's sense of identity that plantations kept growing rice even after the slaves were emancipated and competition from growers in Texas and Louisiana made it less than lucrative to do so.

At Husk and McCrady's, we try to bring the pantry of the Carolina rice kitchen, as Glenn Roberts calls it, back into today's light. Our breads are spiked with benne and rice flour (rice bread being a mainstay of the Colonial Charleston table), and we experiment with everything from rice wines to aged rice. It is an old cuisine reinterpreted—not so much a culinary relic, but an attempt to recapture the heirloom aspects of a complex food culture.

CHARLESTON ICE CREAM

Gullah people of West African origin often cooked rice using a one-pot method known as "soaked rice," where the rice is simmered covered with twice its volume of water for about twelve minutes and then left to sit without lifting the lid for at least another fifteen or twenty minutes. Some Charlestonians call this the "no-peek" style, and it works well with most commercial brands of long-grain rice. True Carolina Gold rice takes a good bit more effort, but the result is the subtle flavors of the rice at their finest. We serve Carolina Gold simply, in bowls, with a dollop of good butter, scattered with herbs and flowers that are in season at the time. Because of its creamy texture, this has been called "Charleston ice cream" for hundreds of years—plus, it can be scooped.

2 teaspoons kosher salt

¼ teaspoon freshly ground white pepper

1 fresh bay leaf

1 cup Anson Mills Carolina Gold Rice (see Resources, page 326)

4 tablespoons unsalted butter, cubed

Herbs, flowers, and benne seeds (optional)

Preheat the oven to 300°F.

Put 4 cups water, the salt, pepper, and bay leaf in a medium heavy-bottomed saucepan and bring to a boil over medium-high heat. Stir to be sure the salt has dissolved, then reduce the heat to medium. Add the rice, stir once, and bring to a simmer. Simmer gently, uncovered, stirring occasionally, until the rice is al dente, about 8 minutes. Drain.

Spread the rice out on a rimmed baking sheet. Dry the rice in the oven, stirring it occasionally, for 10 minutes. Scatter the butter evenly over the rice and continue stirring every few minutes. The rice should be dry in approximately 5 minutes more: all excess moisture should have evaporated and the grains should be separate. Serve immediately with a sprinkle of herbs, flowers, and benne seeds, if you wish.

EINKORN BISCUITS

Einkorn is technically wheat, but the variety is so old that it can behave very differently from the bag of bleached flour you might pick up in the supermarket. In botanical terms, einkorn is very closely related to the first wild forages from which all wheat is derived and, because of that, it's very hard to grow on a large scale: it simply doesn't want to be cultivated. I like that about einkorn. Wild plants sometimes want to stay that way, and the struggle adds intrigue to the flavor.

¾ cup Rendered Fresh Lard (page 316)

6 tablespoons unsalted butter

4½ cups Anson Mills Einkorn Flour (see Resources, page 326)

1 tablespoon plus ½ teaspoon baking powder

1½ teaspoons kosher salt

12 ounces very cold whole-milk buttermilk

Preheat the oven to 425°F. Line a baking sheet with parchment paper.

Freeze the lard and butter for 20 minutes, or until they are solid enough to grate on the large holes of a box grater. Grate them and return to the freezer.

Combine the flour, baking powder, and salt in a large bowl and mix well. Work the frozen lard and butter into the flour with your fingertips until the pieces of butter and lard are about the size of peas. Add the buttermilk and stir until the dough comes together.

Turn the dough out onto a floured work surface. Use a lightly floured rolling pin to roll it out to a ½-inch thickness. Cut out the biscuits with a 2-inch round cutter, flouring it between cuts if necessary, and place them on the prepared baking sheet. The biscuits should just barely touch each other.

Bake the biscuits for 15 to 20 minutes, until golden brown. Remove them from the baking sheet and serve.

NOTE

The biscuits are best served immediately, but they can be made up to 6 hours ahead and reheated briefly in a 350°F oven.

RICE GRIDDLE CAKES

The inspiration for this recipe came from a book called *What Mrs. Fisher Knows about Old Southern Cooking,* written in 1881. Many believe it to be the first African-American cookbook. Mrs. Fisher was a former slave from Mobile, Alabama, and she offered up some incredible recipes, like Calf's Head Stew and Lamb Vigareets. I particularly like this one because it makes use of leftover rice from dinner the night before, a tradition that many can appreciate. These cakes were often served for breakfast, but I eat them any time of day.

½ cup Anson Mills Carolina Gold Rice (see Resources, page 326)

1 cup cornmeal, preferably Anson Mills Antebellum Fine White Cornmeal (see Resources, page 326)

1 teaspoon baking powder

1 teaspoon kosher salt

1 cup whole-milk buttermilk

3 large eggs, lightly beaten

1 tablespoon Rendered Fresh Lard (page 316), melted

4 tablespoons unsalted butter or lard, plus more if needed

Bring 3 cups water to a boil in a medium heavy-bottomed saucepan over medium-high heat. Reduce the heat to medium, add the rice, stir once, and bring to a simmer. Simmer gently, uncovered, stirring occasionally, until the rice is overcooked and a little mushy, about 15 minutes. Drain.

Combine the cornmeal, baking powder, and salt in a small bowl, then stir the mixture into the hot rice. Add the buttermilk, eggs, and melted lard.

Heat the butter on a griddle pan, preferably cast iron, over medium-high heat. Cooking in batches if necessary, spoon the batter onto the griddle to make cakes about 1 inch in diameter. When you see craters in the tops of the cakes, after about 2 minutes, flip them and cook for about 2 minutes on the other side, or until golden and crispy on both sides. (If working in batches, wipe out the pan with paper towels between batches if the fat scorches and add fresh fat.) Serve the griddle cakes immediately.

WHEAT THINS

Wheat thins are almost foolproof and a perfect vehicle for tasting different types of wheat. Anson Mills sells all different kinds of Colonial-period wheats, and they each have different characteristics. I suggest ordering several types of wheat flour and experimenting with them in this recipe.

Wheat thins are the perfect platform for a swipe of Pimento Cheese (page 248).

4 cups Anson Mills Artisan Whole Grain Wheat Flour (see Resources, page 326)

2 tablespoons plus 2 teaspoons sugar

1 tablespoon kosher salt

1 teaspoon baking powder

5 tablespoons cold unsalted butter

5 tablespoons cold Rendered Fresh Lard (page 316)

1 cup boiling water

Position the racks in the upper and lower thirds of the oven and preheat the oven to 350°F. Line two baking sheets with parchment paper.

Combine the flour, sugar, salt, and baking powder in a food processor and pulse to mix well. Add the butter and lard and pulse 10 times. With the processor running, slowly add the water. The dough will be tacky.

Transfer the dough to a floured work surface and lightly knead it into a ball. Using a lightly floured rolling pin, roll it out into a square about ⅛ inch thick. Using a knife or a pizza cutter, cut the dough into 2-inch squares. Prick them with a fork. Arrange the wheat thins on the prepared baking sheets. You will have only a few scraps because of the square cuts, but the scraps can be pressed together, rerolled, and cut again.

Bake the crackers for about 20 minutes, until brown and crispy; halfway through, rotate the sheets top to bottom and front to back to ensure even baking. Remove the wheat thins to a wire rack to cool.

The wheat thins will keep for up to 4 days in an airtight container.

OLD-FASHIONED SODA CRACKERS

MAKES ABOUT
4 DOZEN CRACKERS

A great soda cracker should shatter when you bite into it. It should be thin and crispy, like the industrial saltine crackers available at any old grocer, but when you make these crackers using fresh lard and smoky bacon fat, your opinion of the lowly cracker will change. And your clam chowder may never be the same. I use raw milk, which adds extra fat and flavor to the crackers, but you can substitute whole milk, if you'd like.

2¼ cups Anson Mills Einkorn Flour (see Resources, page 326)

½ teaspoon baking soda

½ teaspoon kosher salt, plus more for sprinkling

¼ cup cold Rendered Fresh Lard (page 316)

¼ cup cold bacon fat

½ cup whole milk, preferably raw

Position the racks in the upper and lower thirds of the oven and preheat the oven to 375°F. Line two baking sheets with parchment paper.

Combine the flour, baking soda, and salt in a medium bowl and mix well. Cut in the lard and bacon fat with two forks until the pieces are the size of peas. Make a well in the center of the mixture, add the milk, and stir well to combine. The dough should look smooth.

Turn the dough out onto a floured work surface and knead it until firm and no longer sticky, about 2 minutes. Using a lightly floured rolling pin, roll the dough out into a square about ⅛ inch thick. Using a knife or a pizza cutter, cut the dough into 2-inch squares. Prick them with a fork and sprinkle lightly with salt. Arrange the soda crackers on the prepared baking sheets. You will have only a few scraps because of the square cuts, but the scraps can be pressed together, rerolled, and cut again.

Bake the crackers for 20 to 25 minutes, or until golden brown; halfway through, rotate the sheets top to bottom and front to back to ensure even baking. Remove the crackers to a wire rack to cool.

The crackers are best eaten when really fresh, but they can be kept for up to 2 days in an airtight container.

SAVORY BENNE WAFERS

MAKES 7 TO 8
DOZEN WAFERS

Benne wafers are synonymous with Charleston, and loads of them leave Market Street every day in the hands of tourists. Unfortunately, most of those benne wafers are made with modern forms of sesame seeds, not the older West African benne, which has a much different flavor. Older forms of benne are harder to grow than newer varieties, but they have a much more savory and floral character, with a beautiful bitterness at the end, giving these wafers an incredible depth of flavor.

1 cup Anson Mills Antebellum Benne Seeds (see Resources, page 326)

2 cups Anson Mills French Mediterranean White Bread Flour (see Resources, page 326)

1 cup Anson Mills New Crop Heirloom Bennecake Flour (see Resources, page 326)

1 teaspoon kosher salt

½ teaspoon baking powder

½ teaspoon baking soda

⅔ cup cold Rendered Fresh Lard (page 316)

⅔ cup cold whole milk

Fleur de sel

Position the racks in the upper and lower thirds of the oven and preheat the oven to 425°F. Line two baking sheets with silicone baking mats or parchment paper.

Toast the benne seeds in a heavy skillet over medium heat, watching carefully and stirring occasionally, until light brown and fragrant, about 5 minutes. Spread them out on a baking sheet to cool.

Sift the bread flour, bennecake flour, salt, baking powder, and baking soda into a large bowl and mix well. Cut the lard in with two forks until the pieces are the size of peas. Fold in the benne seeds. Stir in the milk.

Turn the dough out onto a lightly floured work surface and knead until it comes together, about 3 minutes. Divide it into 3 parts. Pat each part into a disk, then wrap 2 pieces loosely in plastic wrap and set aside. (You do not need to refrigerate them unless your kitchen is very hot.) Using a lightly floured rolling pin, roll the remaining piece of dough out on the floured surface into a paper-thin circle. Prick it all over with a fork and cut out rounds with a 2-inch cutter. Transfer the rounds to the prepared baking sheets and sprinkle them lightly with fleur de sel. Repeat with the remaining dough. Press together any scraps, reroll them paper-thin, and cut out more rounds.

Bake the wafers for about 8 minutes, until golden brown; halfway through, rotate the sheets top to bottom and front to back to ensure even baking. Remove the wafers to a wire rack to cool.

The wafers will keep for up to 5 days in an airtight container.

BENNE-BUTTERMILK ROLLS

MAKES 44 ROLLS

I could eat my weight in these tasty little rolls. We serve them every day at Husk. They are the first things that hit the table when guests arrive, so they have to be special. Bread should make everyone feel comfortable before a meal starts, whether it's in a restaurant or at home.

I like the food at Husk to tell a story, so we make these using Blé Marquis flour, which is a specialty wheat flour from Anson Mills. I want our guests to taste how older varieties of wheat pack so much more flavor. You can substitute all-purpose flour, if you must. A sprinkling of crunchy salt and benne seeds at the end makes the rolls irresistible.

¼ cup sugar

¼ cup local honey

1 teaspoon kosher salt

3 cups all-purpose bread flour

1½ cups all-purpose flour, preferably Blé Marquis flour (see Resources, page 326)

¼ cup crumbled fresh yeast

2 cups whole-milk buttermilk

1 large egg

1 tablespoon water

2 tablespoons Anson Mills Antebellum Benne Seeds (see Resources, page 326)

Fleur de sel

Make a paste with the sugar, honey, and salt in a large bowl. Add both the flours and stir them in with a wooden spoon. In a small bowl, mix the yeast with the buttermilk, then add this to the flour mixture all at once and stir in.

Turn the dough out onto a lightly floured work surface and knead it until smooth, 5 to 6 minutes.

Lightly spray a large bowl with a nonstick spray and place the dough in the bowl. Cover the bowl with a kitchen towel and put it in a warm place. Let the dough rise until it has doubled in size, about 1 hour.

Remove the towel and gently punch the dough down. Cover the bowl again with the kitchen towel and let the dough rise again until doubled in size, about 45 minutes.

Spray two round 9-inch cast-iron skillets with nonstick spray. Portion the dough into 1-ounce rolls: divide the dough in half and then in half again, and divide each portion into 11 pieces. Roll each piece into a ball and carefully place 22 rolls in each pan; they should fit snugly. Cover the pans lightly with kitchen towels, put them in a warm place, and let the rolls rise until they have doubled in size, about 2 hours.

About 20 minutes before the rolls have finished rising, preheat the oven to 400°F.

Whisk together the egg and water to make an egg wash. Using a pastry brush, lightly brush the tops of the rolls with the wash. Sprinkle the rolls with the benne seeds and lightly with fleur de sel. Bake the rolls for about 25 minutes, rotating the pans once halfway through. Test the rolls using a thermometer: the internal temperature should read 195°F. Cool the rolls in the pans on a rack.

The rolls are best served as soon as they have cooled, but they can be kept covered in the pans for up to 1 day and reheated in a 350°F oven for 10 minutes.

FARROTTO
WITH ACORN SQUASH
AND RED RUSSIAN
KALE

SERVES 6
AS A SIDE DISH
OR 4 AS AN ENTRÉE

Like einkorn, farro is an early form of wheat that can be traced back to the Mediterranean, where it was first domesticated. Actually, the word *farro* can refer to the hulled grains of einkorn as well as to emmer wheat and spelt. Spelt is the largest variety, what the Italians call *farro grande,* and it's what I often source from Anson Mills. I like farro for its nutty flavor and chewy texture. If you have ever made risotto, this recipe will look familiar; it uses the same approach—hence the name. Here I pair farrotto with fall flavors, but it can be a vehicle for whatever looks great at the farmers' market. Just keep in mind that farrotto brings a bit more heft to the plate than a traditional risotto.

ACORN SQUASH

1 small acorn squash (about 2½ pounds)

1 tablespoon unsalted butter

1½ teaspoons kosher salt

Scant 1 teaspoon freshly ground white pepper

1 cup Vegetable Stock (page 316)

1 bunch Red Russian or other kale (about 3 pounds)

FARROTTO

2 quarts Vegetable Stock (page 316)

1½ teaspoons canola oil

1½ cups Anson Mills Farro Verde (see Resources, page 326)

4 tablespoons unsalted butter

½ cup diced white onion

1 garlic clove, sliced paper-thin

½ cup dry white wine

1 cup freshly grated Parmigiano-Reggiano cheese

FOR THE SQUASH: Preheat the oven to 425°F.

Cut the squash in half. Remove and discard the seeds and rinse the squash under cold water. Place the squash cut side up on a rimmed baking sheet. Divide the butter between the two halves and sprinkle with the salt and white pepper. Roast the squash for 35 minutes, or until fork-tender.

WHILE THE SQUASH IS ROASTING, PREPARE THE KALE: Remove the stems and ribs from the leaves. Make stacks of the leaves, roll them into cylinders, and cut them into very thin ribbons. Wash the kale in a sink of cold water, changing it several times, to remove any sand. Drain and dry with paper towels.

When the squash is done, remove it from the oven and pour the butter and juices from the cavities into a container. Let the squash cool enough to handle.

Warm the stock in a small saucepan over medium-high heat. Add the butter and juices from the cavities of the squash.

Peel the squash. Place the pulp in a blender, add the warm stock and juices, and blend on high to a very smooth puree, about 3 minutes. Set aside.

FOR THE FARROTTO: Preheat the oven to 425°F.

Heat the stock in a partially covered large saucepan over medium heat; keep warm over low heat.

Heat the canola oil in a medium ovenproof skillet over medium heat. When the oil is shimmering, add the farro and stir to coat it with the oil. Place the skillet in the oven and toast the farro for 8 minutes, stirring after 4 minutes. Transfer the farro to a bowl and reserve. Wipe the skillet clean.

Put the empty skillet over medium-high heat. When the skillet is hot, about 2 minutes, add 2 tablespoons of the butter and reduce the heat to medium. Add the onion and cook, stirring occasionally, until translucent, about 4 minutes. Add the garlic and cook, stirring constantly, until soft, about 2 minutes. Add the wine, increase the heat to high, and cook until the wine is almost evaporated, about 2 minutes.

Add the toasted farro to the pan and stir to coat. Add ½ cup of the warm stock, reduce the heat to medium, and stir until the liquid is almost absorbed. Continue cooking, adding ½ cup of warm stock at a time, stirring to prevent scorching and letting each addition be absorbed before adding the next, until the farro grains have expanded and are al dente, about 1 hour. The farro will look creamy like risotto. *(The farro can be made up to 3 days ahead. Cool to room temperature, cover, and refrigerate. Reheat over low heat before proceeding.)*

TO COMPLETE: Remove the farro from the heat, add the squash puree and kale, and stir until the kale is wilted. Put the skillet back over medium heat and add the remaining 2 tablespoons of butter and the cheese. Stir and serve.

THE YARD

THE HENHOUSE

If you asked any kid what animals live on a farm, I'd bet most would come up with chickens or roosters in their top three. Yet very few large commercial farms today have a single chicken or turkey roaming around, pecking grubs and weed seeds from the pasture. It's also safe to say that unless you go out of your way to find pasture-raised heirloom fowl, the chicken you've been eating doesn't come from what you'd describe as a farm at all. Today the majority of chickens are raised in an environment more closely resembling a factory than a farm.

Most of this is due to the fast-food and prepared-foods industries. The rising popularity of "convenience foods" long ago destroyed the old ways of raising chickens, but I think it's time we brought them back. I don't know about you, but I don't relish the thought of eating birds that have been crammed together by the thousands and are waddling around in their own poop. I'm willing to pay more for a higher-quality bird that has been raised outdoors, in the open air, and that thus tastes remarkably better. Chickens should be raised in a sustainable flock—one that naturally reproduces. Older generations of farmers raised poultry this way, and the results gave cooks many more options for cooking delicious poultry. Today many people are trying to do this at home, and backyard chicken coops have become all the rage.

On a traditional farm, eggs and meat birds were often used as additional sources of revenue. The farmer's family enjoyed fresh chicken meat and eggs and sold the surplus in town. Farmers also allowed a few birds that went "broody" to sit on fertilized eggs and produce the next generation of chicks. And so it went, in a virtuous cycle. And in this continuous cycle, birds could be taken for meat at any age, and that resulted in great diversity in how and when chickens came to market. These days you won't find more than a couple of types of birds in a regular grocery store. Large poultry producers have streamlined production, and the birds themselves, to fit into an assembly-line process. Birds are hatched by the millions, sent out to "growers,"

usually killed at about the same age, and marketed
mostly as "broiler/fryers." In fact, the birds are so
genetically altered that they can't survive much
longer than the six or eight weeks they're allowed
to—their growth rate would outpace their organs and
their underdeveloped legs would fail under their
increasing carcass weight.

In the old days, people raised both broilers and
fryers, and they were distinctly different. They also
raised poussins, roasters, stewing hens, capons,
and roosters; each type had its own advantages
and drawbacks. But the differences were not in
breed or brand name—they were in sex and age.
One backyard flock can produce everything from a
delicious one-pound poussin good for skewering and
grilling to an old rooster best when bobbing in the
famous red-wine-braise preparation that the French
call coq au vin. We've all but lost these traditions in
America, and with it, we've forgotten how to cook
with heritage breeds as well.

As you're cooking the recipes in this chapter,
keep in mind that some may not work as well with
supermarket chicken. Those chickens simply may
not have the age, the flavor, and other characteristics
that allow them to stand up to some methods of
cookery. I encourage you to seek out a farmer, or even
a backyard grower, who can supply you with a few
examples of different types of chicken. It is well worth
the effort and expense.

POULTRY BY CATEGORY

This list of birds is ordered from young, small, and tender to older, large, tough, and most flavorful, with the adult males at the end.

POUSSIN

Some people call these birds spring chickens, and they go by other names as well, but *poussin* is the French term for a young chicken about a month old. They weigh in at about 1 pound each and because they are so amazingly tender, they make great candidates for quick high-heat cooking methods, such as skewering and cooking over an open fire.

BROILER

These chickens come to market at 7 to 12 weeks of age and usually weigh between $1\frac{1}{2}$ and 2 pounds. As their name suggests, they are perfect for the grill. Don't be fooled by supermarket labels; the "broiler" you get in plastic wrap will not compare to a farm-raised specimen.

FRYER

Most commercial chickens in the grocery store are sold as young "broiler/fryers." Although from the right grower, a real fryer, which is probably a bit more mature than the supermarket bird, can stand the direct high heat of a charcoal fire, many traditional breeds will seem tough to those accustomed to tender factory-raised chicken. What you'll get in return in a farm-raised traditional bird is flavor. And when soaked in buttermilk and fried, it'll turn anyone into a barnyard evangelist.

ROASTER

By modern poultry standards, roasters mature to a ripe old age. These old girls tip the scales at up to 8 pounds and can be up to a year old. That's geriatric territory for a meat chicken, but the firm muscling of an older bird adds a depth of flavor to a dish that younger birds can never match.

STEWING HEN

Traditional farmers don't tend to waste much, and they're not going to keep feeding a chicken that isn't going to produce. So when an old layer's production declines, it's off to the stew pot. If you want to make real chicken stock, this bird is the ticket. Stewing hens are cheap, and they are versatile. The traditional chicken and dumplings should also be made with this bird.

CAPON

A capon is a castrated cockerel, and a cockerel is a young rooster. Some people object to the castration process, but tasting a high-quality capon might help them get over that detail. Caponizing a bird changes its hormonal balance and allows it to fatten into a real butterball. A capon is double the size of a rooster, fat and delicious, and its meat more tender than that of a hen. It's a royal bird, the product of a technique that dates to the Roman Empire.

COCK

South Carolina likes its cocks. They even named the University of South Carolina's athletic teams after them. But those are fighting cocks, which, although still battled illegally in the state, aren't too good to eat. Old roosters, the kind that stay around the farmstead to service hens, are few and far between. They're noisy and obnoxious and can be aggressive. Roosters are old and sinewy; the meat is tough and can be gamy. But paired with assertive flavors and cooked ever so slowly, they are the basis for some legendary preparations.

FRIED CHICKEN AND GRAVY

(OR THE WAY I MAKE FRIED CHICKEN AT HOME)

SERVES 2

I've worked on my fried chicken for many years, researching every recipe that I could lay my hands on, from early antebellum instructions to the Kentucky Colonel's secret technique. This recipe uses five fats, and each one contributes to the flavor of the result.

To do the chicken right, you need an old black cast-iron skillet with a lid. Sure, you can make it in a deep fryer (like we do at the restaurant), but I prefer the old-fashioned way, which is nearly impossible to pull off in a restaurant. The skillets take up so much stove space that you can't make more than ten orders at a time. So this isn't the fried chicken you're going to eat at Husk. This is the way grandmas cook fried chicken in the South, and it's the way everyone should be making fried chicken at home.

This recipe takes a lot of time and attention, way more than most conventional approaches (the chicken must be brined for 12 hours, so plan ahead). But it's good. Be sure to ask your butcher for the chicken skins to render for fat and to save the cooking fat, which makes mighty fine gravy. I've thrown that recipe in here too, to complete the meal just like my grandma would have.

BRINE

1 gallon water

38 regular black tea bags or 4 ounces loose tea, such as Charleston Plantation Tea (see Resources, page 326)

1 cup kosher salt

1 cup sugar

1 chicken (about 3 pounds)

2 quarts whole-milk buttermilk

3 tablespoons Husk Hot Sauce (page 238)

1 tablespoon plus 1 teaspoon freshly ground black pepper

1½ pounds chicken skins cut into ½-inch squares

6 cups flour, preferably Anson Mills White May Flour (see Resources, page 326)

1 cup fine cornmeal, preferably Anson Mills Antebellum Fine White Cornmeal (see Resources, page 326)

2 tablespoons cornstarch

1½ teaspoons garlic powder

1½ teaspoons onion powder

½ teaspoon cayenne pepper

½ teaspoon smoked paprika (see Resources, page 326)

1 cup Rendered Fresh Lard (page 316)

1 cup canola oil

2 ounces Benton's slab bacon (see Resources, page 326), diced

2 ounces Benton's smoked ham (see Resources, page 326), diced

2 tablespoons unsalted butter

Sea salt

Gravy (recipe follows)

FOR THE BRINE: Put the water in a pot and bring to a boil over high heat. Remove from the stove, add the tea bags, and let them steep for 8 minutes.

Remove the tea bags, or strain the liquid if you used loose tea. Add the salt and sugar to the hot water and stir to dissolve them. Pour the brine into a heatproof container and cool it to room temperature, then refrigerate until completely cold.

Cut the chicken into 8 pieces: 2 legs, 2 thighs, 2 wings, and 2 breast pieces. Rinse with cold water. Place in the brine, cover, and refrigerate for 12 hours.

After the chicken has spent 12 hours in the brine, make an ice bath in a large bowl with equal amounts of ice and water. Place the chicken in the ice bath for 5 minutes. (The ice will rinse away any impurities.) Remove the chicken and pat it dry.

Combine the buttermilk, hot sauce, and 1 tablespoon of the black pepper in a large container. Add the chicken to the buttermilk mixture, cover, and let marinate for 1 hour at room temperature.

While the chicken is marinating, put the chicken skins in a small saucepan over very low heat, adding a small amount of water to prevent the skins from sticking and burning. Cook the skins, stirring frequently so that they don't burn, until their fat is rendered. Strain the fat; you need 1 cup.

Drain the chicken, quickly rinse under cold water, and pat dry.

Combine the flour, cornmeal, cornstarch, garlic powder, onion powder, the remaining 1 teaspoon black pepper, the cayenne pepper, and smoked paprika in a large bowl and mix well. Add the chicken and toss to coat thoroughly. Allow it to sit for 15 minutes, then shake off any excess, transfer the chicken to a wire rack, and let sit for 15 minutes.

Meanwhile, put the chicken fat, lard, and canola oil in a large, deep cast-iron skillet. Add the bacon and ham and heat the fats to 275°F. Turn the heat off and allow the bacon and ham to infuse the fats and oil for 10 minutes.

With a skimmer or slotted spoon, remove the bacon and ham from the skillet (discard them or eat as a snack) and heat the oil to 300°F. Add the breasts and thighs and cook for 3 minutes. Add the legs and wings and cook for 5 minutes. (Remove the fat needed for the gravy at this point and start the gravy; see recipe on page 102.)

Turn the chicken over, cover the skillet, and cook until the pieces of chicken are the color of hay, about another 5 minutes. Remove the lid, turn the pieces again, cover, and cook the chicken until golden brown, another 3 minutes. Add the butter and continue cooking, turning the pieces once, for another 2 minutes or so on each side. The chicken should be crispy and golden brown. Let the chicken rest and drain on wire racks or on a plate covered with paper towels for about 8 minutes, but no longer.

Sprinkle with sea salt and serve with the gravy.

<hr>

NOTE

If you use a large, deep cast-iron skillet and the recommended 3-pound chicken, a small bird called a fryer, you shouldn't have any trouble frying all the chicken at one time. If that isn't possible, use two skillets and mix two batches of fat to achieve the flavor and crispness imparted by the combination of fats.

GRAVY

<hr>

MAKES ABOUT 2½ CUPS

¾ cup plus 1 tablespoon cooking fat from the fried chicken

¾ cup all-purpose flour

2 cups whole milk

1 tablespoon kosher salt

1 tablespoon freshly ground black pepper

1 tablespoon soy sauce

After the chicken has fried for 8 minutes and is ready to turn, carefully remove the fat needed to make the gravy from the skillet. Put ¾ cup of the fat in a medium saucepan over medium heat, stir in the flour to make a roux, and cook for 2 minutes, stirring constantly. Gradually whisk in the milk. Reduce the heat to medium-low and cook the gravy, stirring occasionally to be sure that you don't have any lumps, for 15 minutes to cook out the taste of the raw flour.

Add the salt, pepper, and soy sauce, then whisk in the remaining tablespoon of cooking fat to make the gravy glossy. The gravy can be kept warm, covered, over the lowest heat for up to 20 minutes.

CRISPY FRIED FARM EGGS
with FRESH CHEESE, PICKLED CHANTERELLES, WILD WATERCRESS, and RED-EYE VINAIGRETTE

SERVES 6

I first saw an egg prepared this way at my friend Donald Link's Herbsaint in New Orleans. They serve an egg that has been poached and then fried on top of spaghetti. Sounds odd, tastes unbelievable. Later I had a similar version while visiting Blue Hill at Stone Barns. Other versions are popping up all over the place, and I love the concept. The egg arrives at the table golden brown and crispy and when it's cut into, the yolk spills out and sauces the plate.

In this salad, we combine that effect with a simple technique for making your own cheese at home. Try making the cheese with different kinds of milk, perhaps goat or sheep. The red-eye vinaigrette was inspired by red-eye gravy, which is always popular in the South, probably because coffee and ham taste really good together, especially with eggs! Here I lighten up the traditional flavors. The vinaigrette works perfectly in this dish, but it also tastes great on roasted fish.

Note that the cheese must drain overnight.

CHEESE
1 quart whole milk

½ cup heavy cream

2½ tablespoons white vinegar

¼ teaspoon kosher salt

⅛ teaspoon freshly ground white pepper

VINAIGRETTE
1¼ cups cider vinegar

¼ cup honey vinegar, such as Jean Marc Honey Vinegar (see Resources, page 326)

½ cup plus 2 tablespoons rendered ham fat, warmed

1¼ cups grapeseed oil

1 tablespoon sugar

1 tablespoon fresh lemon juice

1 tablespoon instant coffee granules

CRISPY EGGS
Kosher salt

11 large eggs

2 quarts canola oil

2 cups all-purpose flour

4 cups panko bread crumbs, finely ground in a food processor

Freshly ground black pepper

SALAD
12 cups (about 1 pound) watercress, preferably wild

1 small red onion, shaved as thin as possible into rings

1½ cups Pickled Mushrooms (page 214), preferably chanterelles, drained

FOR THE CHEESE: Line a strainer with cheesecloth and place it over a bowl. Heat the milk and cream in a large nonreactive saucepan over high heat until the temperature reaches 170°F; use a digital instant-read thermometer to check. Add the vinegar, salt, and white pepper and keep the mixture at 170°F for 2 minutes. Curds should form. Pour the mixture into the strainer, place the bowl and strainer in the refrigerator, and refrigerate overnight.

The next day, transfer the cheese to a container and refrigerate until ready to use. *(Any leftovers can be kept in a tightly covered container for up to 3 days in the refrigerator.)* Reserve the whey that drained into the bowl for another use, such as cooking grits or making bread or pancakes. *(Tightly covered, the whey will keep for up to 5 days in the refrigerator or 3 months in the freezer.)*

FOR THE VINAIGRETTE: Combine the vinegars in a bowl and then slowly whisk in the remaining ingredients. This is a broken vinaigrette and won't be smooth. *(Tightly covered, the vinaigrette can be kept for up to 3 days in the refrigerator. Bring to room temperature before using.)*

FOR THE EGGS: Fill a medium saucepan with water and bring it to a boil over high heat; add 2 tablespoons salt. Make an ice bath in a medium bowl with equal amounts of water and ice. Place 6 of the eggs in the boiling water and boil for exactly 5 minutes and 15 seconds. Transfer the eggs to the ice bath. Just as soon as they are cool enough for you to handle them, peel the eggs while still in the ice bath, then remove and drain well.

Put the canola oil in a fryer or large pot and heat it to 350°F. Meanwhile, make a breading station by putting the flour and bread crumbs in separate shallow bowls and seasoning with salt and pepper. Whisk the remaining 5 eggs with a splash of water in a shallow bowl.

Working in two batches, dredge the boiled eggs in the flour, gently shaking off the excess, dip them into the whisked eggs, allowing any excess to fall back into the bowl, and dredge them in the bread crumbs, gently shaking off any excess. Fry them until golden brown, about 2 minutes; the yolks should still be fluid when the eggs are cut into. Drain them on paper towels and use at once.

TO COMPLETE: Dress the watercress and onions with the vinaigrette to your taste. Divide them among six plates. Crumble 2 tablespoons of cheese over each salad and top with an egg and 4 or 5 pickled mushrooms.

GRILLED CHICKEN WINGS WITH BURNT-SCALLION BARBEQUE SAUCE

MAKES 12 PIECES

I am borderline obsessed with chicken wings. It's the perfect food after a long work shift or on a chill day with your friends, crushin' cheap American beers in the backyard. It's food that allows you to let your guard down. After all, you're eating food cooked on the bone with your hands and licking the sauce from your fingers in between chugs of ice-cold beer. Pure heaven.

Note that the wings must be brined overnight.

BRINE

8 cups water

¼ cup kosher salt

1 tablespoon sorghum (see Resources, page 326)

WINGS

6 chicken wings, cut into tips and drumettes

3 tablespoons green peanut oil (see Resources, page 326)

1 tablespoon Husk BBQ Rub (page 311)

¾ cup thinly sliced scallions (white and green in equal parts)

½ cup dry-roasted peanuts, preferably Virginia peanuts, chopped

SAUCE

10 scallions, trimmed

1 tablespoon peanut oil

Kosher salt

1 cup Husk BBQ Sauce (page 236)

1 tablespoon Bourbon Barrel Foods Bluegrass Soy Sauce (see Resources, page 326)

1 cup cilantro leaves

EQUIPMENT

1 pound hickory chips

Charcoal chimney starter

3 pounds hardwood charcoal

Kettle grill

FOR THE BRINE: Combine the ingredients for the brine. I brine the wings using either a heavy-duty plastic bag that the wing tips can't puncture or a Cryovac machine (you use a lot less brine this way). Place the wings in the brine and turn to cover well. Refrigerate overnight.

Soak the wood chips in water for a minimum of 30 minutes but preferably overnight.

FOR THE SAUCE: Toss the scallions in the peanut oil and season with salt. Lay them out on the grill rack and heavily char them on one side, about 8 minutes (the charred side should be black). Remove them from the grill and cool for about 5 minutes. Clean the grill rack if necessary.

Put the scallions and the remaining sauce ingredients in a blender and process until smooth, about 3 minutes. Set aside at room temperature.

FOR THE WINGS: Fill a chimney starter with 3 pounds hardwood charcoal, ignite the charcoal, and allow to burn until the coals are evenly lit and glowing. Distribute the coals in an even layer in the bottom of a kettle grill. Place the grill rack as close to the coals as possible.

Drain the wings; discard the brine. Dry the wings with paper towels, toss in the peanut oil, and season with the BBQ rub.

Place the wings in a single layer on the grill rack over the hot coals and grill until they don't stick to the rack anymore, about 5 minutes. Turn the wings over and grill for 8 minutes more. Transfer the wings to a baking sheet.

Drain the wood chips. Lift the rack from the grill and push the coals to one side. Place the wood chips on the coals and replace the rack. After about 2 minutes, place the wings in a single layer over the side of the grill where there are no coals. Place the lid on the grill, with the lid's vents slightly open; the vents on the bottom of the grill should stay closed. Smoke the wings for 10 minutes. It's important to monitor the airflow of the grill: keeping the lid's vents slightly open allows a nice steady flow of subtle smoke.

Remove the wings from the grill, toss them in the sauce, and place them on a platter or in a serving pan. Top with the chopped scallions and peanuts and serve.

QUAIL STUFFED WITH MORELS AND CORNBREAD, WITH BAKED RED PEAS AND GREEN GARLIC PUREE

SERVES 6

Quail graces many a table in the South. I like to serve it stuffed, because the stuffing helps to keep the bird moist while cooking and the meat is less likely to overcook. This stuffing with fresh morels is one that I often make for Thanksgiving. The baked red peas were inspired by BBQ baked beans. It only seems right to serve them alongside an iconic bird like the bobwhite quail. Both speak to the long history of hunting, gathering, and gardening so precious to the South.

STUFFING

3 cups kosher salt

1 pound fresh morels

1 recipe Cracklin' Cornbread (page 71)

½ cup small dice Vidalia onion

½ cup small dice celery

2 tablespoons chopped tarragon

1 teaspoon celery seeds

½ cup Chicken Stock (page 318)

1 large egg, lightly beaten

QUAIL

6 semiboneless quail (about 4 ounces each), rinsed under cold water

Kosher salt and freshly ground black pepper

¼ cup duck fat (see Resources, page 326)

1 bunch thyme

2 tablespoons unsalted butter

4 garlic cloves, smashed and peeled

GREEN GARLIC PUREE

2½ pounds green garlic

1 cup Vegetable Stock (page 316)

1 tablespoon cream cheese

Baked Sea Island Red Peas (page 54), warm

FOR THE STUFFING: To clean the morels, fill a large, clean bucket or deep container with 5 gallons warm water, add 1 cup of the salt, and stir until it is dissolved. Add the morels and let them soak for 1 hour.

Using a wire rack, push the morels down a little and, with your other hand, skim any leaves and debris off the top with a mesh strainer. Remove the rack and gently lift out the morels, being careful not to disturb the debris that falls to the bottom. Repeat this procedure twice using the remaining 2 cups salt, then lay the morels out on a wire rack and let them air-dry at room temperature for 30 minutes to 1 hour, until they are completely dry. Slice the morels into ¼-inch rounds.

Crumble the cornbread into a bowl, add the sliced morels, onion, celery, tarragon, celery seeds, chicken stock, and egg, and stir well. The stuffing should be evenly moist.

FOR THE QUAIL: Stuff each quail with about ½ cup of the stuffing. You will have some left over, which you can bake (in a 350°F oven) or freeze for later use. Season the quail with salt and pepper.

Heat two large cast-iron skillets over very high heat. Add 2 tablespoons of the duck fat to each skillet. When the fat shimmers, add 3 quail to each skillet, breast side down, and cook until golden brown, about 5 minutes. Remove the quail from the skillet, reduce the heat to medium, and arrange half the thyme in each skillet to make a bed for the quail. Place the quail seared side up on the thyme and divide the butter and garlic between the skillets. Cook, basting the quail, until they are cooked through, about 6 minutes. The quail are best served at once, but you can hold them in a 200°F oven for up to 10 minutes.

MEANWHILE, FOR THE GREEN GARLIC PUREE: Shave the garlic, equal parts green and white, as thin as possible on a mandoline. Wash the shavings in several changes of water.

Place the garlic and stock in a large saucepan, bring to a simmer over medium-high heat, and simmer until the garlic is tender, about 7 minutes. Blend the vegetable stock and garlic in a blender on high until very smooth, about 5 minutes. Add the cream cheese and blend for another 2 minutes.

TO COMPLETE: Pour ¼ cup of the garlic puree into the center of each of six warm plates. Place the baked peas on the puree and top each plate with a quail.

CHICKEN SIMPLY ROASTED IN A SKILLET

SERVES 2
HUNGRY PEOPLE

I love cooking chicken like this at home: it fills the house with an amazing aroma. The flavors of garlic, lemon, and parsley are classic and simply delicious. You can throw the dish together at the last minute with minimal shopping and prep. I serve it with a simple salad or a very fresh vegetable quickly cooked on the side.

GARLIC CONFIT

6 large garlic cloves, peeled

1 teaspoon sugar

1 tablespoon kosher salt

1 teaspoon freshly cracked black pepper

2 tablespoons extra-virgin olive oil

CHICKEN

1 whole chicken (about 3 pounds)

Kosher salt and freshly cracked black pepper

½ cup canola oil

PAN SAUCE

2 cups Chicken Stock (page 318)

1 tablespoon all-purpose flour

1 cup flat-leaf parsley leaves cut into very thin strips

Grated zest (use a Microplane) and juice of 1 lemon

FOR THE GARLIC CONFIT: Preheat the oven to 400°F. Cut two 12-inch squares of aluminum foil and lay one piece on top of the other. Place the garlic cloves on the foil. Sprinkle with the sugar, salt, and pepper. Pour the olive oil over the garlic cloves. Shape the foil into a pouch by bringing the edges of the foil together over the garlic and sealing them. Flatten the bottom of the pouch so it will stay upright in the oven and place it on a baking sheet.

Roast the garlic for about 30 minutes, until the cloves are very soft but not falling apart. Set the garlic aside in the pouch. Leave the oven on.

MEANWHILE, FOR THE CHICKEN: Using kitchen shears, cut down along both sides of the backbone, then clip it out. Cut the wings off at the first joint. (Freeze the backbone and wing tips to make stock.) Split the chicken in half. Use paper towels to dry the skin. Season both sides of the chicken with salt and pepper. Place the chicken in a baking dish and let it sit at room temperature for 20 minutes.

Place two 12-inch cast-iron skillets over high heat. When the skillets smoke, add ¼ cup of the canola oil to each. As soon as the oil smokes, carefully add a half chicken to each skillet, skin side down. Weight each chicken half with another heavy skillet or pan so it stays flat and browns evenly. Cook the chicken, with the weights on it, until the skin is crispy and evenly browned, 5 to 7 minutes. Remove the weights.

Flip the chicken over, and place the skillets in the oven. Roast the chicken for about 20 minutes, until an instant-read thermometer inserted into the thickest part of the thigh reads 155°F. Place the chicken on plates to rest while you make the sauce.

FOR THE PAN SAUCE: Combine the roasting juices and fats from both skillets into one; set aside. Place the other skillet on the stove over medium heat until it is hot to the touch, about 1 minute. Pour 1 cup of the chicken stock into the skillet and use a spatula to scrape the browned bits from the bottom of the skillet, then gently boil the stock to reduce it by half, about 5 minutes. Add the remaining cup of stock and set aside.

Place the skillet with the roasting juices over medium heat. Sprinkle the flour evenly over the juices and gently whisk it in until there are no lumps. Reduce the heat to low and cook for 2 minutes, stirring constantly with the whisk; do not let the roux get too dark around the edge. Whisk in the chicken stock, making sure to fully emulsify it. Increase the heat to high and bring the sauce to a simmer, then reduce the heat to medium-high and reduce the sauce until it coats the back of a spoon, about 5 minutes.

Add the parsley, lemon zest and juice, and 2 tablespoons of the garlic oil from the pouch of garlic and whisk to combine.

TO COMPLETE: Place the garlic and pan sauce over the chicken and enjoy.

BREAST OF GUINEA HEN PAN-ROASTED ON THE BONE

with OATS, RAMPS, and PARSNIPS

SERVES 4

Guinea hens, native to Africa, thrive in the low brush and scrub around a farm. They are noisy little birds, but their dark meat sure is delicious. Here I slow cook the breasts and serve them alongside seasonal vegetables and heirloom oats. Oats have great significance in the Low-country. They appear in many old recipes but were mostly eaten for breakfast. For this dish I wanted a savory, almost risotto-like porridge to go along with the roast guinea hen. The oats can be served year-round, studded with other seasonal vegetables. This recipe makes use of parsnips and the wild ramps that appear in early spring.

OATS

3 tablespoons unsalted butter

1 leek, white and green parts, cut in half, thinly sliced, and washed well

1 large garlic clove, thinly sliced

1 cup dry white wine

1½ cups Anson Mills Brewster Oats (see Resources, page 326)

5 cups Chicken Stock (page 318)

1 teaspoon kosher salt

½ teaspoon freshly ground white pepper

1 cup freshly grated Pecorino Romano cheese

RAMPS AND PARSNIPS

3 tablespoons unsalted butter

2 large parsnips (about 1½ pounds), peeled, halved lengthwise, any hard core removed, and cut into ½-inch-thick slices

1 teaspoon kosher salt

½ cup Vegetable Stock (page 316)

8 ramps, cleaned and hairy root ends removed

½ teaspoon fresh lemon juice

GUINEA HENS

4 bone-in guinea hen breasts

Kosher salt and freshly ground black pepper

About 2 tablespoons canola oil

3 tablespoons unsalted butter

1 garlic clove, lightly smashed and peeled

15 thyme sprigs

Black Truffle Jus (page 322)

FOR THE OATS: Melt 2 tablespoons of the butter in a large saucepan over medium heat. When the butter is foaming, add the leek and cook, stirring occasionally, until softened, about 5 minutes. Add the garlic and cook, stirring occasionally, until softened, about 3 minutes more. Add the white wine, stir to combine, and cook until almost all the wine has evaporated, about 7 minutes.

Stir in the oats and cook, stirring frequently, for 1 minute. Add 1 cup of the stock, bring to a simmer, and simmer, stirring frequently, until it is nearly absorbed by the oats. Continue adding 1 cup of stock at a time and cooking, stirring frequently, until the liquid is nearly absorbed after each addition; the oats are done when they are tender with the slightest chewiness, about 25 minutes. Add the salt and white pepper, remove the saucepan from the stove, and stir in the cheese and the remaining tablespoon of butter. You can keep the oats warm, covered, over very low heat for up to 10 minutes.

WHILE THE OATS COOK, MAKE THE RAMPS AND PARSNIPS: Heat 2 tablespoons of the butter in a large saucepan over medium heat. When the butter is foaming, add the parsnips, season with the salt, and cook over medium heat, stirring occasionally, until softened, about 5 minutes. Add the vegetable stock and ramps and cook until the ramp bulbs are soft, about 2 minutes. Stir in the remaining tablespoon of butter and the lemon juice. You can keep the vegetables warm, covered, over very low heat for up to 10 minutes.

FOR THE GUINEA HENS: Heat a large cast-iron skillet over high heat. Add enough oil to cover the bottom of the skillet. When the oil shimmers, add the breasts skin side down, and sear them for 2 minutes without shaking the skillet or touching the breasts. Reduce the heat to medium and cook, checking the underside of the breasts frequently, until they are golden brown, about 4 minutes (8 minutes if you refrigerated them). Flip the breasts over and add the butter, garlic, and thyme to the skillet. Cook, using a large spoon to baste the breasts with the butter, until an instant-read thermometer inserted in the fattest part registers 158°F, about 8 minutes. Transfer the breasts to paper towels to blot any fat, then cut the meat away from the bone and blot it again with paper towels.

TO COMPLETE: In a small saucepan, warm the Black Truffle Jus over medium heat, about 3 minutes.

Place ⅔ cup of the oats in the center of each of four large, warm plates. Drizzle the jus around the oats. Place one guinea hen breast on top of the oats on each plate. Divide the ramps and parsnips among the plates, scattering them over the breasts.

TENNESSEE FOIE GRAS with COUNTRY HAM, STRAWBERRY–MEYER LEMON JAM, AND HEIRLOOM JOHNNYCAKES

SERVES 4
AS AN APPETIZER

Foie gras and country ham are ingredients near and dear to my heart. I get foie gras from Specialty Duck Farm in Hohenwald, Tennessee. It's great having a foie gras producer in the South that raises its animals with care. There's a trend of serving seared foie gras with a healthy dose of great finishing salt and that gave me the idea for this. Instead of using salt, I garnish the plates with some wonderful smoky country ham from my friend Allan Benton. Allan has become quite the legend among cured-pork aficionados. He has a way with hickory wood like no one else, and I can always pick his products out in a blind tasting. The ham gives the dish a more Southern personality. I complement these flavors with a jam from our larder and some really delicious johnnycakes.

JOHNNYCAKES

2 cups cornmeal, preferably Anson Mills Antebellum Fine Yellow Cornmeal (see Resources, page 326)

½ cup Anson Mills Artisan Whole Grain Wheat Flour (see Resources, page 326)

1 teaspoon kosher salt

1 teaspoon baking powder

1 teaspoon baking soda

1 large egg

1¾ cups whole-milk buttermilk

FOIE GRAS

Four 2-ounce slices of fresh Grade A foie gras (see Resources, page 326)

Kosher salt and freshly ground black pepper

About ¼ cup Strawberry–Meyer Lemon Jam (page 216)

4 thin slices Benton's country ham (see Resources, page 326), about 3 inches by 3 inches, at room temperature

1 teaspoon finely sliced chives

FOR THE JOHNNYCAKES: Mix the cornmeal, flour, salt, baking powder, and baking soda in a medium bowl. Whisk together the egg and buttermilk in a small bowl, then whisk the wet ingredients into the dry ingredients. *(The batter can be covered and refrigerated for up to 12 hours.)*

FOR THE FOIE GRAS: Preheat the oven to 200°F.

Using a sharp knife, score each piece of foie gras in a ¹⁄₁₆-inch-deep diamond pattern all over the surface. Season generously with salt and pepper.

Select a cast-iron skillet large enough to easily hold the 4 pieces of foie gras and heat it over medium-high heat until very hot. Add the foie gras and brown on each side, 1½ to 2 minutes per side. It should still feel soft all the way through. Transfer the foie gras to a rimmed baking sheet and place in the oven to keep warm.

Remove half of the rendered fat from the skillet and set it aside. Reduce the heat to medium. Working in batches, spoon the johnnycake batter in the size of half dollars into the pan and cook the cakes until they are golden brown on the bottom, about 2 minutes. Flip the cakes and cook the other side until golden brown, 1 to 2 minutes more. Transfer the cooked cakes to another rimmed baking sheet and keep warm in the oven while you cook more johnnycakes. When you have used about half the batter, wipe out the skillet, then add the reserved foie gras fat, heat the fat, and continue cooking johnnycakes.

TO COMPLETE: Place a small dollop of jam on each plate. Place a piece of foie gras beside the jam and arrange a stack of 3 johnnycakes beside the foie gras. Garnish each plate with a slice of country ham, and sprinkle the foie gras with chives.

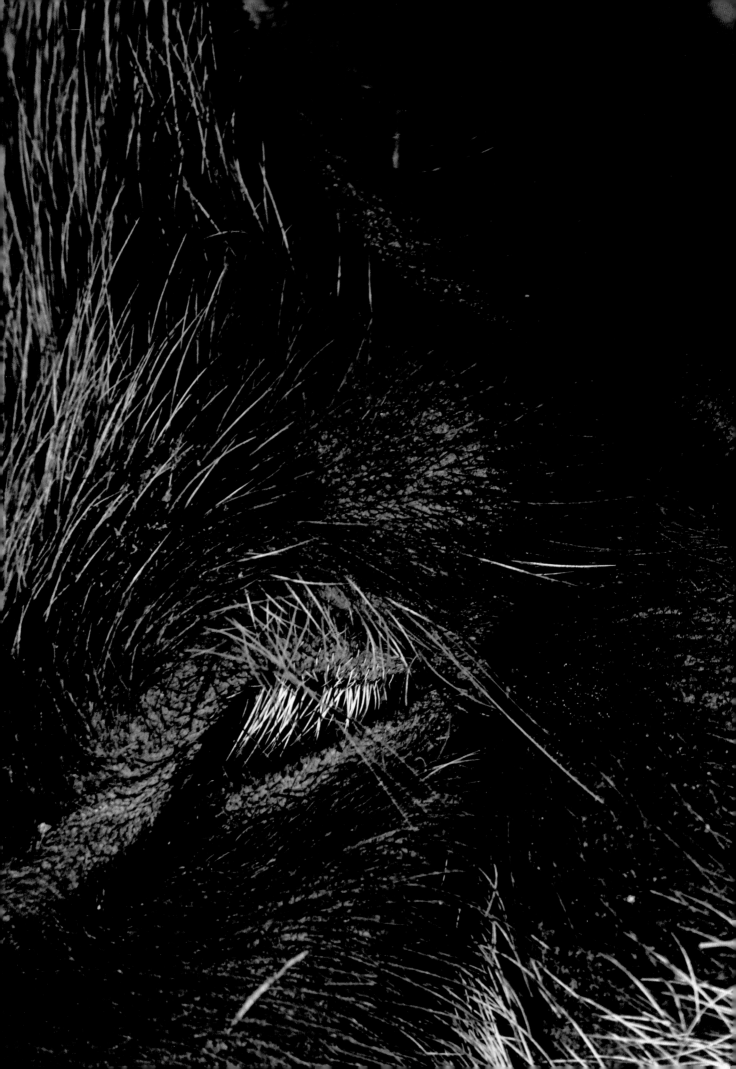

THE PASTURE

HIGH ON THE HOG

After creating a market garden to serve the kitchens of Husk and McCrady's, and realizing how profoundly the experience changed my respect for food, the next logical step was raising animals. But that's a huge responsibility. Something changes inside you when you're tending animals that pump blood and breathe in the air and you know they'll end up on a cutting board in the kitchen and then on a plate in the dining room. It may sound silly, but when you become attached to an animal, feed it, worry about its safety and health, play games with it, and watch it grow, that beast becomes a small piece of you.

I've kept quite a few pigs at my friend Shawn Thackeray's farm. I love to feed them old milk from the local dairy. After a while, they get so excited about this treat that they see me coming and trot over toward my truck, then surround me and jostle for attention. Some of them let me pour the milk right into their mouths. But even pigs have likes and dislikes. Mine love chocolate milk and hate strawberry milk. We (the pigs and I) hold taste tests of vegetables and fruits, and they let me know which ones they like the best, sorting through the piles of vegetables and eating their favorite things first, leaving the prickly cucumbers or bitter greens for the last bites. But have no doubt: in the world of a pig, nothing goes uneaten. On my days off, I like to buy a six-pack of beer and just hang out watching them eat, watching them search through the dirt and forage for things to snack on. For me, the sound of pigs munching on acorns is magical, and watching a young gilt trot around with half a pumpkin stuck on her head is priceless.

But these encounters are not all warm and fuzzy. I can taste the complexity in future country hams every time I watch a pig root around for acorns. Eventually the day will come when that pig has to meet its maker, which undoubtedly brings one of the most difficult of burdens. Corralling a pig off six acres into a trailer is no easy task. We've improved our technique over time, but I can testify that no amount of milk, acorns, porridge, leftover pizza, or even a long trail of Froot Loops can persuade a smart pig to walk willingly to the gallows.

Dropping off your first pigs at the abbatoir can be a humbling experience. After the first time I did it, I couldn't sleep that night. But when you pick up the carcass of beautiful pork raised in a humane and responsible manner, there is a sense of pride involved. To be a chef means to buy and cook meat, and that means we have a choice to make. The differences between the animals that modern agribusinesses produce and animals raised on pasture and humanely treated cannot be understated. Commercial animals are treated horribly, given inferior feed and no attention, and confined in huge warehouses where their feet never touch the dirt. If horses were treated this way, someone would get arrested—but I'm sure you all have seen plenty of good documentaries and read informative books to this effect.

My relationships with people who raise animals properly extend to sheep and cattle. Craig Rogers is my only source for lamb. I've been to his farm. I see the work he does and the care he puts into raising his animals. Bear Creek Farm in Tennessee and Double H Farms are my two sources for beef. I want to use only breeds that are suitable for our region. Understanding the connection between the animals and the land means raising animals the right way.

My goal is to create a demand to keep heritage breeds alive. I believe that eating these animals, those of heritage breed and of proper husbandry, guarantees their very survival. Unless someone cares enough to take the extra effort and lays aside a bit of profit in the pursuit of better culinary outcomes, these breeds will simply cease to exist. No one will grow them without a ready demand. By pursuing rare breeds, supporting sustainable methods of production, and demanding a product in line with our ethics, we ensure that great ingredients don't fall victim to the industrialized holocaust of modern agribusiness. If we need to learn just one lesson, it's that patience pays off.

HERITAGE
LIVESTOCK BREEDS

No run-of-the-mill pastured animals here; these are my favorite breeds.

KATHADIN LAMB

When most people think of lamb, they picture big, wooly balls of fleece. In the South, we often grow "hair sheep," of which the Kathadin is a particularly delicious example. Unlike northern sheep, breeds of hair sheep, which originated in Africa, shed their pelts in the spring. They are also more naturally resistant to parasites. And I think they are more delicious. Many people object to the gamy flavor some lamb can have, but the meat of hair sheep tastes relatively milder. The supposed drawback? They produce smaller cuts than conventional lamb—but I'll take great flavor over quantity any day.

OSSABAW HOGS

The mere presence of the Ossabaw hog in America is something of a miracle. It's a strange breed, having been isolated on its namesake island off the coast of Georgia for more than two centuries before being "rediscovered" by scientists, who used it for diabetes research. The Ossabaw originally came from Spain in the 1600s, with explorers from Iberia. They often dropped off some swine on barrier islands knowing that it would create a meat source for future travelers, but the Ossabaw remained relatively untouched, enough to develop insular dwarfism and the peculiar ability to pack on incredible amounts of fat during the acorn season on the islands. To see an Ossabaw today is to look back into history; they are a true form of old pork, with long snouts, a nasty demeanor, and bristling black Mohawks that grow all the way down their backs. But what we are after is the fat—the intramuscular fat of the Ossabaw breed makes it one of the most desirable.

GUINEA HOGS

Like the Ossabaw, the Guinea hog is a product of its historical environment. No one really knows where it originated, but a hundred years ago, these were some of the best hogs one could stock a small homestead with. They are self-sufficient, can eat most anything, and provide a ready source of manure—they're a garbage disposal at one end and a fertilizer wagon at the other. They'll even keep the barnyard free of snakes and other critters, which they'll hunt down if given the chance. But most important, they pile on loads of fat, and in contrast to the insipid, tasteless fat of a commercial hog, the lard of a Guinea hog (and the Ossabaw too), when raised on a natural diet of vegetable scraps and acorns, takes on a luscious flavor. For someone with a curing closet and a smokehouse, it's a beautiful taste to behold. Cured properly, the Guinea hog's lard melts into pure porky bliss the moment it hits your tongue.

PINEYWOODS CATTLE

The heritage Pineywoods cattle that used to roam the barrens of the Southeast were almost completely wiped out in the twentieth century. Before farmers enclosed their cattle with fencing, there were great longleaf pine forests that supported a vast ecosystem that's now virtually extinct. Pineywoods flourished in this environment, and the task of rounding them up for slaughter gave birth to America's first cowboys, many of them slaves or former slaves. Like the Ossabaw hog, Pineywoods cattle were left along the southern U.S. coast by Spanish explorers. Over time, they developed a robust resistance to common cattle diseases and parasites and a knack for avoiding predators, as well as the ability to subsist on rough vegetation that today's domestic cattle would find completely unpalatable. Because of this, I think Pineywoods are some of the best "grass-fed" stock around, even if their lack of a grain diet means they don't get to spend a life among bucolic green pastures.

SLOW-COOKED RIB EYE
WITH POTATO CONFIT AND GREEN GARLIC–PARSLEY BUTTER

SERVES 6

I know, I know, meat and potatoes . . . so avant-garde. But sometimes one exceptional meal at home with loved ones can be just as special as a twenty-course tasting menu at a grand restaurant.

Slow-cooking is a technique that lends itself well to a large cut like the rib eye. The secret is twofold: get a good sear on the meat before placing it in the oven, and arrange it so that the delicious fat cap slowly bastes the meat as it cooks.

Such a decadent cut of meat topped with a flavorful pat of butter deserves a sinful side dish, and this potato confit certainly fills the bill. It can be made well in advance and stored in the fridge. In fact, the longer it sits, the better it tastes; the potatoes just continue to soak up all that tasty and delicious fat.

GREEN GARLIC–PARSLEY BUTTER

- 2 cups chopped green garlic (green and white parts)
- 1 pound unsalted butter, slightly softened
- 1 cup chopped flat-leaf parsley
- 1 cup minced shallots
- Grated zest of ½ lemon (use a Microplane)
- ¼ cup fresh lemon juice
- 2 tablespoons Worcestershire sauce
- 1 tablespoon plus 1 teaspoon kosher salt
- 2 teaspoons freshly ground black pepper
- 2 teaspoons anchovy paste

RIB EYE

- 1 center-cut bone-in rib-eye roast (about 7.5 pounds), deckle and fat cap left on
- Kosher salt and freshly ground black pepper
- Canola oil
- 15 thyme sprigs
- 15 rosemary sprigs
- 1 garlic bulb, cut in half
- 5 cups Heirloom Potato Confit (page 46)

FOR THE GREEN GARLIC–PARSLEY BUTTER: Bring a large pot of salted water to a boil. Make an ice bath in a bowl with equal parts ice and water. Put the green garlic in a strainer and submerge it in the boiling water for 7 seconds, then remove and submerge it in the ice bath until completely cold. Remove from the ice bath, shake off the excess water, then drain and dry on paper towels.

Put the green garlic in a blender and blend on high until smooth, about 5 minutes; add a splash of water as needed to keep the blade running smoothly.

Combine the garlic puree, butter, parsley, shallots, lemon zest, lemon juice, Worcestershire sauce, salt, pepper, and anchovy paste in the bowl of a stand mixer fitted with the paddle attachment and mix on low speed until thoroughly blended, about 2 minutes. Divide the butter in half and put each portion on a sheet of plastic wrap. Roll each one into a log and wrap tightly in the plastic. Place in the freezer and freeze until solid.

FOR THE RIB EYE: Preheat the oven to 250°F. Place a rack in a roasting pan.

Liberally season the beef with salt and pepper. Heat a large cast-iron skillet over high heat. When the skillet is hot, add ¼ inch of canola oil. When the oil begins to smoke, add the beef fat side down and sear until golden brown, 3 to 5 minutes. Repeat on all sides. Remove from the heat.

Cover the rack in the roasting pan with the thyme, rosemary, and garlic. Place the beef on the herbs and garlic bulb halves, fat side up. Put the pan in the oven and roast the beef for about 2 hours and 45 minutes, until the internal temperature reaches 125°F. Remove the pan from the oven and let the beef rest for 25 to 30 minutes before carving it. Baste the beef with the pan juices several times as it rests. Remove the green garlic butter from the freezer 1 hour before serving.

TO COMPLETE: Carve the rib eye into 6 slices and arrange on warmed plates. Top each slice with a ½-inch-thick disk of room-temperature green garlic butter and serve with the potato confit.

=== NOTE ===

This recipe makes more green garlic–parsley butter than you will need for the rib eye, but it can be frozen, tightly wrapped, for up to 1 month and used in other dishes.

HUSK CHEESEBURGER

MAKES 10 CHEESEBURGERS

When I opened Husk, I knew we had to have a cheeseburger on the menu. Everyone has their own idea of the perfect burger; mine was inspired by the drive-in that my family used to take me to when I was young. Robo's was the only real "restaurant" in my hometown, and my family just loved it. We would go there after my Little League baseball games. As a game wound down, I would be daydreaming about that burger, shake, and crinkle-cut fries. It's probably the reason for some missed fly balls.

What I remember most about the cheeseburger was the squishy bun and how wonderful it was to eat the double patty covered in gooey American cheese. This recipe is a tip of the hat to that burger. I've changed it a little to make it my own—I wouldn't dare try and replicate the burger from Robo's. This recipe feeds a crowd, but you can halve it for a smaller group.

If you don't have a meat grinder, ask the butcher to grind the meats for you.

SPECIAL SAUCE

1¾ cups mayonnaise, preferably Duke's (see Resources, page 326)

1¼ cups yellow mustard

5 tablespoons ketchup

½ cup Bread-and-Butter Pickles (page 228), drained and cut into ⅛-inch dice

¼ cup pickled jalapeños, drained and cut into ⅛-inch dice

Grated zest (use a Microplane) and juice of 1 lemon

1 tablespoon Husk Hot Sauce (page 238)

Kosher salt and freshly ground black pepper to taste

2 tablespoons pepper vinegar, preferably Texas Pete brand

CHEESEBURGERS

One 3-pound fresh boneless chuck roast

12 ounces fresh flank steak

3 ounces bacon, preferably Benton's (see Resources, page 326)

3 tablespoons unsalted butter, at room temperature

10 hamburger buns, preferably potato rolls

1 cup shaved white onion

20 slices American cheese

50 Bread-and-Butter Pickles (see page 228)

FOR THE SAUCE: Combine all of the ingredients in a large container and stir together to blend well. Cover, and refrigerate. *(Tightly covered, the sauce will keep for up to 5 days in the refrigerator.)*

FOR THE CHEESEBURGERS: Grind the chuck, flank steak, and bacon through a meat grinder fitted with the large die into a bowl. Mix gently to combine. Then run half of the mixture through the small die. Mix the two together.

Portion the meat mixture into twenty 3-ounce patties, about ½ inch thick (each burger gets 2 patties). If not cooking right away, arrange on a baking sheet, cover tightly with plastic wrap, and refrigerate. *(The patties can be refrigerated for up to 1 day. Remove from the refrigerator about 30 minutes before you're ready to cook; it's important that the patties are not ice-cold when they hit the hot pan.)*

Generously butter the tops and bottoms of the buns. Toast on a griddle until nice and golden brown. Reserve.

Heat two 12-inch cast-iron skillets until as hot as possible. Divide the patties between the two hot pans. When the patties are nice and charred, about 2 minutes, flip them over and cook for 2 minutes more for medium. Place the onion slices on 10 of the patties. Place a slice of the cheese on all of the patties and allow it to melt, about 30 seconds. Stack the non-onion patties on top of the onion patties. Remove from the heat.

Smear both sides of the buns with special sauce. Place 5 pickles on the bottom half of each bun. Add the burger patties and top with the top halves of the buns. Serve at once.

CHARRED BEEF SHORT RIBS WITH GLAZED CARROTS AND BLACK TRUFFLE PUREE

SERVES 4

This is a poor man's meat combined with a rich man's treasure—unless you are lucky enough to have truffles growing in your backyard. My friend Tom Michaels grows Périgord truffles at his Tennessee home, and they were the inspiration for this dish. When they are in season, they are worthy of a splurge.

You can get a great effect from combining the tenderness of an old-school braise with the char flavor from the grill pan. And once you master this braising technique, which works with a variety of tougher meats, you'll never be intimidated by braising again.

You will need to order the boneless short rib from a specialty butcher. Note that the braised short rib must be refrigerated overnight.

SHORT RIBS

One 1-liter bottle Cabernet Sauvignon

4 cups Chicken Stock (page 318)

1 slab boneless chuck short ribs (about 4 pounds), cleaned of any connective tissue

Kosher salt and freshly ground black pepper

2 cups large dice sweet onions

1½ cups ½-inch-thick carrot slices

4 celery ribs, leaves discarded, cut into 1-inch pieces

6 garlic cloves, smashed and peeled

1 orange, cut into quarters

1 Bosc pear, cut into quarters

8 thyme sprigs

1 fresh bay leaf

BLACK TRUFFLE PUREE

1 golf-ball-sized Tennessee Black Périgord Truffle (see Resources, page 326)

1 cup Périgord Black Truffle Juice (see Resources, page 326)

2 tablespoons red wine vinegar, such as Banyuls (see Resources, page 326)

1 tablespoon soy sauce

1 teaspoon kosher salt

½ cup grapeseed oil

½ recipe Carrots Braised and Glazed in Carrot Juice (page 46), warm

Carrot tops for garnish

FOR THE RIBS: Preheat the oven to 350°F.

Heat a large heavy roasting pan over high heat on the stovetop. Add the red wine, bring to a boil, and cook until reduced by half, about 10 minutes.

Reduce the heat to medium-high, add the chicken stock, and bring to a low boil. Cook until reduced to about 3 cups, 15 to 20 minutes. Remove from the heat.

Season the short ribs with salt and pepper. Put a rack in the roasting pan (with the wine and chicken stock still in it), and place the short ribs on the rack. Cover the pan, put it in the oven, and cook the short ribs for 1½ hours.

Remove the roasting pan from the oven. Leaving the beef on the rack, scatter the onions, carrots, celery, garlic, orange, pear, thyme, and bay leaf around it. Place the uncovered pan back in the oven and continue to cook until the meat is tender but not falling apart, about 1 hour.

Carefully remove the short ribs and set aside in a baking pan, to catch any juices. Strain the braising liquid into a bowl; discard the solids. Remove any fat from the top. Allow to cool on the countertop, skimming additional fat as it rises to the top; this will take a couple of hours.

When all of the fat has been removed from the braising liquid, pour it and any juices released from the short ribs over the meat, cover, and refrigerate overnight. The short ribs will soak up some of the braising liquid and remain moist.

The next day, place a roasting pan over high heat. Remove the short ribs from their pan and pour the braising liquid into the clean roasting pan; return the short ribs to the refrigerator. Bring the braising liquid to a simmer, reduce the heat to medium-high, and cook, skimming any fat, until the jus reduces to coat the back of a spoon, about 20 minutes. Strain the jus, return it to the pan, and keep warm over low heat.

Heat a large cast-iron grill pan over very high heat. Season the short ribs with salt and pepper. When the grill pan is extremely hot, add the beef and char it for about 8 minutes on each side. Transfer the beef to a cutting board to rest.

When the meat is cool enough to handle, divide it into 4 portions and add them to the reduced jus. Bring the jus to a light simmer and baste the beef as it heats through. (If necessary, the short ribs can be held in a 200°F oven for up to 1 hour.)

MEANWHILE, FOR THE BLACK TRUFFLE PUREE: Chop the truffle and place it in a blender, along with the truffle juice, vinegar, soy sauce, and salt. Blend on high until smooth, about 3 minutes. With the blender running, drizzle in the grapeseed oil. The result will be almost a paste. (The black truffle puree can be kept in a tightly covered container in the refrigerator for up to 5 days. Let come to room temperature before using.)

TO COMPLETE: Drizzle black truffle puree across four plates. Place a piece of beef on each plate and coat with the jus. Place 2 baby carrots beside the beef and glaze the carrots with the carrot juice. Garnish with some carrot tops.

CORNMEAL-FRIED PORK CHOPS

WITH GOAT CHEESE–SMASHED POTATOES AND CUCUMBER AND PICKLED GREEN TOMATO RELISH

SERVES 6

You'd be hard-pressed to find a soul food restaurant that does not serve a fried pork chop. Pork chops are so important to the Southern diet that they come in at least half a dozen cuts, some bone-in, some paper-thin, all delicious. I bread mine in a cornmeal crust and fry them up in plenty of hot grease, and I like to pair them with smashed potatoes. I add goat cheese to the potatoes to fancy them up and serve the dish with a green tomato relish.

Note that the pork chops must marinate overnight.

PORK CHOPS

6 boneless pork chops (about 3 ounces each)

1 quart whole-milk buttermilk

Kosher salt and freshly cracked black pepper

2 cups cornmeal, preferably Anson Mills Antebellum Fine Yellow Cornmeal (see Resources, page 326)

Cayenne pepper

Canola oil for shallow-frying

GOAT CHEESE–SMASHED POTATOES

15 medium Red Bliss potatoes (about 3 pounds), washed

Kosher salt

1 cup half-and-half

6 ounces goat cheese

8 tablespoons (1 stick) unsalted butter, diced and chilled

½ cup finely sliced chives

Freshly ground white pepper

Cucumber and Pickled Green Tomato Relish (page 236), drained

FOR THE PORK CHOPS: Pound each pork chop between two pieces of plastic wrap to ⅛ inch thick. Place the chops in a container and cover them with the buttermilk. Cover and marinate overnight in the refrigerator.

FOR THE SMASHED POTATOES: Put the potatoes in a large pot, cover with water, and add ¼ cup salt. Bring the water to a simmer over medium-high heat, reduce the heat to medium-low, and cook the potatoes until fork-tender, about 30 minutes; do not let the water boil.

Preheat the oven to 200°F.

Meanwhile, remove the chops from the buttermilk (discard it) and season them with salt and cracked pepper. Put the cornmeal in a shallow bowl and season it with salt and cayenne pepper. Dredge the chops in the cornmeal, gently shaking off the excess, and put on a large plate.

Heat two large cast-iron skillets over high heat. When the skillets are hot, add ¼ inch of canola oil to each and heat for 1 minute. Carefully place 2 pork chops in each skillet; do not shake the skillets or touch the chops for 1 minute. Then reduce the heat to medium-high and cook the chops until golden brown, about 4 minutes. Carefully turn the chops over and cook until golden brown and crispy on both sides, about 4 minutes more. Transfer the chops to a baking sheet and keep warm in the oven. Discard the oil in one skillet, replace it with new oil, and heat the oil over high heat. Cook the remaining 2 pork chops in the same way; transfer to the oven to keep warm.

When the potatoes are almost cooked, bring the half-and-half to a simmer in a small saucepan over medium heat. Drain the potatoes and place them in a large bowl. Using a wooden spoon, carefully smash each potato without breaking it apart. Pour the hot half-and-half over the potatoes; crumble the goat cheese and the butter over them, and fold in the chives. You should have small pockets of goat cheese throughout and the potatoes shouldn't be mashed. Season the potatoes with white pepper.

TO COMPLETE: Divide the pork chops and potatoes among six warm plates. Spoon some relish on top of the chops.

SLOW-COOKED PORK SHOULDER WITH TOMATO GRAVY, CREAMED CORN, AND ROASTED BABY VIDALIAS

SERVES 12

Cooking a large joint of pork reminds me of home—it's a meal of memories. The smell of a pork shoulder slow-roasting in the oven—this one roasts a long time—is pretty close to heaven, and that first little piece of the pork crust right out of the oven, when no one is looking, is the ultimate reward. The mustardy spice rub caramelizes into a perfect foil for the tender, fatty meat.

Tomato gravy is something that my mom taught me to make when I was a little kid. She would stir the cornmeal roux until it was the perfect consistency. My love of onions comes from my dad. We went camping a lot when I was a kid, and if we were camping, we were fishing. My favorite part of the campfire meal was when my dad would wrap onions in aluminum foil and bury them in the fire. Opening the foil and revealing the meltingly tender onions was always something I looked forward to, whether there was fish or not.

Note that the pork shoulder cooks for about 14 hours, and plan accordingly.

PORK

2 tablespoons brown sugar

2 tablespoons kosher salt

2 tablespoons freshly ground black pepper

1 tablespoon paprika

1 bone-in pork shoulder (also called butt; about 6 pounds), skin removed

½ cup Dijon mustard

ONIONS

6 baby Vidalia onions with greens attached (about 8 ounces each)

2 tablespoons unsalted butter

6 thyme sprigs

3 garlic cloves, lightly smashed and peeled

½ teaspoon kosher salt, plus more to taste

TOMATO GRAVY

2 tablespoons bacon fat

2 tablespoons cornmeal, preferably Anson Mills Antebellum Fine White Cornmeal (see Resources, page 326)

3 cups home-canned tomatoes or canned San Marzano tomatoes

1 tablespoon kosher salt

1 tablespoon freshly cracked black pepper

A triple recipe of Creamed Corn (page 49), warmed

FOR THE PORK: Preheat the oven to 250°F. Place a rack in a roasting pan.

Combine the brown sugar, salt, pepper, and paprika in a small bowl and blend well. Using a brush, paint the top only of the pork with the mustard. Pat on the seasoning mixture. Place the pork on the rack in the pan and roast, uncovered, for about 14 hours, until the meat is tender but not falling apart; baste it occasionally with the pan juices during the last hour to make a glaze. Remove the pork from the oven, transfer it to a platter, and let it rest for 10 minutes. Reserve the juices in the roasting pan, skimming off any fat from the top as the pork rests.

ABOUT 2 HOURS BEFORE THE PORK IS DONE, PREPARE THE ONIONS: Remove the greens from the onions, slice the greens as thin as possible, and reserve to use as garnish. Place the butter, thyme, and garlic on a large piece of aluminum foil and top with the onions. Fold up the edges of the foil and seal to make a closed packet. Place the packet in a baking pan. Add the onions to the oven for the last 2 hours of the pork's cooking time.

MEANWHILE, FOR THE TOMATO GRAVY: Heat the bacon fat in a large nonreactive saucepan over high heat. Stir in the cornmeal with a wooden spoon, reduce the heat to low, and cook, stirring constantly, until the cornmeal turns a light brown color, about 5 minutes.

Using your hands, crush the tomatoes into bite-sized pieces, then add the tomatoes and their juices to the pan and stir to combine. Increase the heat to medium, bring the gravy to a simmer, and cook, stirring occasionally, until it is slightly thickened and the cornmeal is soft, about 10 minutes; be careful that it is not sticking or scorching. Add the salt and pepper. Keep warm over low heat for up to 1 hour.

TO COMPLETE: Remove the onions from the oven, carefully open the packet, and cut the onions into quarters. Put the onions in a dish, baste with the liquid left inside the foil, and season with the salt.

Portion the pork by gently pulling it into large chunks with a pair of tongs. Serve with the onions, creamed corn, and tomato gravy. Sprinkle the pork with the reserved onion greens.

HOW TO BUILD A PIT AND COOK A WHOLE PIG LIKE A CHAMPION PITMASTER

Here's a "recipe" to build a whole hog barbeque pit, courtesy of my friend Pat Martin of Martin's Bar-B-Que Joint in Nashville, Tennessee. You can build the pit anywhere and make it temporary or permanent—your choice. Then roast a whole pig the way I do.

Source your pig ahead of time. You can buy a pig direct from a farmer or order one from your local butcher shop. Think about how many people you want to serve. A 160-pound pig is going to feed 160 people. Scale up or down depending on your party size. Ask for the pig to be butterflied.

WHAT YOU'LL NEED FOR THE PIT

100 cinder blocks

Five 5-foot-long pieces ½-inch-thick rebar

A 4-foot level

A measuring tape

One 24-by-4-inch lintel (or angle iron)

One 4½-by-3½-foot sheet of expanded metal or cattle or hog fencing

One 6-by-5-foot piece of roofing tin or sheet metal

Mortar and sand if you're going to build a permanent pit

A cooler full of beer

The best thing to do, especially if you don't have a truck or trailer, is to have Home Depot or Lowe's deliver the materials to your location. You're going to work hard enough building the pit and then cooking the hog all day or night, so pay to have the materials delivered to you. You may need to source the roofing tin or sheet metal from a local farm store.

First drink a cold beer. Then find some level ground on which to build your pit.

Mark a 5-by-4-foot rectangle and start laying the cinder blocks down end to end on the perimeter of the rectangle. After you've made one row, stop and drink another beer. Use the level to make sure you're building this thing evenly. Lay the second row on top, making sure to offset the blocks from the first row for stability. Leave out a block or two on each end of this row to create your shovel openings.

Take your mason's lintel and lay it over the open space, basically "bridging" the other two blocks on either side of the empty space. Begin laying the third row, offsetting the blocks again.

Now is a good time to step back and survey your work to make sure everything is going as planned, and you might as well have another beer. Then lay the fourth row the same way.

When the fourth row is finished, place the rebar evenly across the blocks. Lay the expanded metal or fencing on top. Finally, lay the fifth row of blocks. The purpose of the fifth row is to support the roofing tin that will cover the hog, which will be cooking on the expanded metal.

Sit down and rest. Have a beer and look around for a few bricks or rocks to lay on top of the tin when you get your hog on; it'll keep the wind from blowing up under it.

Note: If you want to make your pit permanent, fill the blocks with sand for added insulation as you build it and add mortar between the blocks.

CREATE A FIRE OR BURN PIT

You'll need a place off to the side to burn your wood down to coals. There are many ways to do this. You can start a good-sized campfire, you can use an old diesel barrel (see picture) or you can use 9 to 10 cinder blocks to make a small burn pit. If you do that, you'll need two or three 2-foot-long pieces of

rebar. Use 5 to 6 blocks for the base row, lay the rebar across, and make another row of blocks on top. Build your fire on top of the rebar.

You'll need at least half a cord of wood (whatever is locally available). The wood that burns best with the best flavor is wood that's been aged for a year. Fire up your wood in your burn pit. As the wood burns down to coals, it will fall through the rebar and you can easily scoop it with a shovel. You'll need to start this fire a good 3 or 4 hours before you plan on putting your hog on. It will take you 2 or 3 hours to get enough coals burned down, then at least another hour getting your pit hot before you start cooking.

ROAST THE PIG

The roasting will take approximately 6 to 12 hours, so set yourself up with the cooler of beer, some snacks, a playlist, and some good friends. Make sure your burn pit is hot with coals. Place a thermometer on the edge of the tin cover.

Season the flesh of the pig generously with Husk BBQ Rub (page 311), making an even coating.

When you begin, to get the pit extra hot, place two shovelfuls of coals under the pig's shoulders and the hams (hind legs) and spread them out. Check your thermometer; you should get the pit up to 250°F for the first hour. Put the pig skin side up on top of the expanded metal. Cover it with the roofing tin. Weight down the cover with some blocks, bricks, or rocks.

After the first hour, flip the pig so it's skin side down. Flipping a pig is a two-man job. From here on out, you want to keep the temperature at a constant 225°F. This is the hard part. Check the temperature every 30 minutes and add more coals every 20 minutes or so under the hams and shoulders.

So, how long does it take to cook? It depends on how big the pig is. A 160-pound pig will cook in 10 to 12 hours; an 80-pound one will cook in about 6 hours. You'll know the pig is done when you can reach in and pull away the rib bones with no resistance, clean as a whistle. The shoulder bone should come out clean too.

When the pig is ready, pull out all the bones and throw them away. Take a knife and break up the meat. Add Husk BBQ Sauce (page 236) and serve warm, right out of the pit.

CRISPY SWEETBREADS

WITH SPICY RED PEPPER GLAZE, EGG, BROCCOLI, AND PUFFED RICE

SERVES 6

I put this dish on the McCrady's menu soon after I started there. We were brainstorming after service, as we often do, and I wanted to talk about sweetbreads, because we were throwing away a whole lot of them. I went on a rant: "Why the hell don't more people order sweetbreads? They're amazing! If we don't figure out a way to sell more, we will have to stop serving them, because they are too expensive to throw in the trash. It's heartbreaking, and I'm tired of walking around with a broken heart over this shit."

So I told my team that we were going to think of sweetbreads as chicken and make a list of America's favorite chicken dishes. My old friend from culinary school, Andy Allen, raised his hand and blurted out, "General Tso's Chicken!" The meeting was adjourned, we ordered takeout for lunch the next day to get the full effect of that famous dish, and the rest is delicious history. Thank you, Andy!

Furikake is a blend of seaweed and dried fish. Togarashi is a Japanese spice mix with chile pepper.

Note that you'll need to start this recipe 2 days ahead. The sweetbreads need to be brined for 24 hours and then compressed for another 24 hours. And you need to dehydrate the rice overnight. It's a long process, but the results are worth it.

SWEETBREADS

1 gallon water

1 cup kosher salt, plus more for seasoning

½ cup sugar

2 fresh bay leaves

1 large lobe veal sweetbreads (about 3 pounds)

4 cups Vegetable Stock (page 316)

1 cup dry white wine

1 garlic clove, crushed and peeled

12 thyme sprigs

2 quarts canola oil or 2¾ cups peanut oil plus 1¼ cups Rendered Fresh Lard (page 316)

3 cups cornstarch

Young shiso leaves

PUFFED RICE

Kosher salt

1 cup Anson Mills Carolina Gold Rice (see Resources, page 326)

2 cups canola oil

2 tablespoons furikake flakes (see Resources, page 326)

SPICY RED PEPPER GLAZE

3 pounds red bell peppers, cored and seeded

½ cup fresh orange juice

1 cup sugar

¾ cup soy sauce

⅔ cup mirin

¼ cup rice wine vinegar

1 tablespoon sambal oelek

1 cup water

2 tablespoons cornstarch

EGG

Egg yolks from 4 fresh eggs

½ teaspoon kosher salt

BROCCOLI

1 small head broccoli (about 12 ounces)

2 tablespoons miso

1 tablespoon unsalted butter, room temperature

Togarashi chile flakes (see Resources, page 326)

Kosher salt

EQUIPMENT

Dehydrator

Juice extractor

Immersion circulator

Vacuum sealer

FOR THE SWEETBREADS: Two days ahead, bring 2 quarts of the water to a boil in a large pot. Add the 1 cup salt, the sugar, and 1 bay leaf and stir to dissolve the salt and sugar. Add the remaining cold water, pour the brine into a heatproof container, and cool to room temperature, then refrigerate until the temperature of the brine has dropped to 40°F.

Put the sweetbreads in the brine, cover, and refrigerate for 24 hours.

The next day, transfer the sweetbreads to a colander (discard the brine) and rinse under cold water for 2 minutes.

Combine the vegetable stock, wine, garlic clove, thyme, and the remaining bay leaf in a nonreactive pot and bring to a boil over high heat. Reduce the heat to medium-low and simmer the poaching liquid for 30 minutes.

Remove the pot from the heat and add the sweetbreads. Cover the pot and allow the sweetbreads to poach in the hot liquid, off the heat, until white and firm, about 30 minutes.

Drain the sweetbreads and place on a rimmed baking sheet. To compress the sweetbreads, weight them down by placing a heavy baking sheet on top of them and a large cast-iron skillet on top of the baking sheet. Refrigerate for 24 hours.

AFTER REFRIGERATING THE SWEETBREADS, START THE RICE: Combine 4 cups water and 2 teaspoons salt in a medium saucepan and bring to a boil over high heat. Add the rice, stir once, and bring to a simmer, then reduce the heat to the lowest possible setting and simmer the rice for 8 minutes, or until tender. Drain the rice in a colander and rinse thoroughly; the water should run clear and the grains should be separate.

Dry the rice overnight in a dehydrator at 180°F, until it is very dry. You should be able to snap the grains in half.

TO PUFF THE RICE: Heat the canola oil to 400°F in a medium saucepan over high heat. Remove from the stove and add 2 tablespoons of the dehydrated rice. The rice will puff and rise to the top. Drain the oil through a fine-mesh strainer into another saucepan. Transfer the rice from the strainer to one of the prepared baking sheets and spread it in a single layer. Sprinkle it with some salt and furikake flakes. Bring the oil in the saucepan back to 400°F and repeat the process. Transfer the rice to two baking sheets lined with paper towels. It will take 4 or 5 times to puff all of the rice. The rice can be held at room temperatue for up to 6 hours.

FOR THE SPICY RED PEPPER GLAZE: Run the peppers through a juice extractor; you need 2½ cups of pepper juice.

Combine the red pepper and orange juices in a medium nonreactive saucepan, bring to a boil over high heat, and reduce by half, about 15 minutes. Skim off any scum on top.

CONTINUED

Add the sugar, soy sauce, mirin, rice wine vinegar, and sambal, bring to a boil, and reduce by half again, about 10 minutes.

Put the water in a small bowl and, using a fork, stir the cornstarch into the water. Slowly whisk this slurry into the saucepan and bring the mixture to a simmer. Remove from the heat and let cool to room temperature, then cover and refrigerate. *(The glaze can be refrigerated for up to 2 days.)*

FOR THE EGG: Preheat the water bath in an immersion circulator to 67°C (see sidebar, page 118). Blend the yolks with a fork in a small bowl and mix in the salt. Place the yolks in a small vacuum bag and seal it. Cook in the water bath at 67°C for 2 hours.

When the yolks are done, make an ice bath in a bowl with equal parts ice and water. Chill the yolks to 39°F in the ice bath.

Cut open the vacuum bag. The yolks should have thickened into a pudding-like texture. Pass the yolks through a fine-mesh sieve and put them in a squirt bottle or piping bag (cut the tip off when ready to pipe the eggs). Refrigerate until ready to use. *(The yolks can be refrigerated for up to 1 day.)*

FOR THE BROCCOLI: Cut the florets from the stem. Reserve the stem and 6 florets. Bring a large pan of salted water to a boil. Put the remaining florets in a strainer and submerge them in the water until tender, about 1 minute. Drain.

Transfer the blanched florets to a blender, add the miso and butter, and blend on high until smooth, about 5 minutes. Set aside. *(The puree can be cooled, covered, and refrigerated for up to 1 day.)*

JUST BEFORE SERVING: Peel the membranes from the sweetbreads and discard. Divide them into 6 portions. Pour the canola oil into a deep fryer or a large pot and heat it to 350°F.

Meanwhile, bring a medium saucepan of salted water to a boil. Peel the broccoli stem with a vegetable peeler and cut it into ½-inch-thick disks.

Put the reserved 6 broccoli florets in a strainer and submerge them in the boiling water until hot, about 20 seconds. Remove and season with togarashi flakes and salt. Do the same with the disks of stem.

Reheat the red pepper glaze over low heat; keep warm. Reheat the broccoli puree over low heat; keep warm.

Put the cornstarch in a large bowl. Season the sweetbreads with salt. Toss the sweetbreads in the cornstarch, coating them liberally. Fry them until crispy, about 5 minutes. Toss them in the warm red pepper glaze.

TO COMPLETE: Smear the broccoli puree onto six warm plates. Add a dollop of the puree to each. Place the sweetbreads on the smear. Arrange 3 broccoli florets and 3 stem slices around the sweetbreads on each plate. Pipe the egg yolk inside the puree. Sprinkle the rice and shiso leaves over the sweetbreads.

GRILLED LAMB HEARTS WITH BUTTER BEAN PUREE, VADOUVAN, AND CORN AND SWEET POTATO LEAVES

SERVES 6

One night while I was savoring some Pappy Van Winkle with Craig Rogers, the conversation drifted to the importance of buying whole animals, and how that is the best deal for both chefs and farmers. He told me about his freezers full of lamb kidneys and hearts. I'd never cooked a lamb heart, so he promised to send me some to experiment with.

When they arrived, I butterflied a couple of them and cleaned them, tossed them in some olive oil, seasoned them, and threw them on the grill for a bit. I let them rest and sliced them up for everyone to taste. The kitchen staff was nervous at first, but in a few minutes, the entire plate was gone. It turns out that lamb hearts are an unheralded delicacy, with a perfect texture, similar to that of the loin, and a very pleasant density.

Sweet potato leaves carry a similar story. I can never understand why more people in the South don't eat them. Other cultures consider them a treat and here are all these farmers in the South growing sweet potatoes who have never tasted the leaves, much less tried to sell them. You may need to seek out the lamb hearts and sweet potato leaves from specialty stores and farmers' markets, but I hope this dish turns people on to two humble ingredients that deserve more of our attention.

CORN AND SWEET POTATO LEAVES

4 ears corn in the husk, soaked in water for 1 hour

12 sweet potato leaves

1 tablespoon Vadouvan Spice (page 309)

1 tablespoon unsalted butter

LAMB HEARTS

6 whole lamb hearts

Kosher salt

½ cup extra-virgin olive oil

2 garlic cloves, thinly shaved

1 tablespoon freshly cracked black pepper

½ cup chopped flat-leaf parsley

¼ cup chopped rosemary

BUTTER BEAN PUREE

2 cups fresh butter beans

2 cups Vegetable Stock (page 316)

2 teaspoons kosher salt

1 tablespoon unsalted butter

FOR THE CORN AND SWEET POTATO LEAVES: Preheat the oven to 400°F.

Put the corn on a rimmed baking sheet and bake it for 15 minutes, or until the kernels are soft. Remove it from the oven.

When the corn is cool enough to handle, remove the husks and silk and cut the kernels from the cobs. Set aside. (This can be done up to 1 day ahead. Refrigerate the kernels, and bring them back to room temperature when ready to proceed.)

FOR THE LAMB HEARTS: Butterfly the lamb hearts lengthwise. Remove any visible tendons or sinew. Season the hearts with salt and let stand at room temperature for 30 minutes.

Combine the olive oil, garlic, pepper, parsley, and rosemary in a large bowl. Add the hearts, tossing to coat. Cover and refrigerate for 30 minutes.

Meanwhile, prepare a very hot grill; allow the grill grates to preheat so that the hearts won't stick.

FOR THE BUTTER BEAN PUREE: Put the butter beans and stock in a medium saucepan, bring to a simmer over high heat, and cook for 3 minutes.

Transfer the mixture to a blender, add the salt, and blend on high until very smooth, about 5 minutes. Add the butter and blend until incorporated. The puree can be held in a covered saucepan on the back of the stove for up to 20 minutes (don't let it get too hot or sit too long, or it will turn brown).

TO FINISH THE CORN AND SWEET POTATO LEAVES: Wash the leaves, drain them, and pat them dry. Make stacks of the leaves, roll them into cylinders, and cut them into ½-inch-wide ribbons.

Combine the corn and sweet potato leaves in a large skillet over medium heat. Add the Vadouvan and butter and toss to coat. (This can be kept warm on the back of the stove for no more than 8 minutes. It should be made as close to serving time as possible.)

Remove the hearts from the marinade and grill them, turning once, over the hottest part of the grill with the grill lid open, for about 2 minutes per side. The hearts are done when an instant-read thermometer reads 140°F. (You don't want the hearts to be more than medium, or they will be tough.)

TO COMPLETE: Place the butter bean puree on each of six large warm plates. Cut each heart into 4 slices. Place the slices on top of the puree and top with the corn and sweet potato leaf mixture.

CRISPY PORK TROTTERS WITH GARLIC SCAPES, ENGLISH PEAS, AND PORCINI

SERVES 6

Whoever came up with the phrase "high on the hog" probably never had a proper fritter made from "the nasty bits": the feet, the skin, the head, the tail, the organs—where the true flavor of the pig lies.

When these fritters are fried, the outside turns super-crispy and the inside just melts. The crispy trotters go with just about anything. Heck, serve them with a simple salad of arugula. I like to serve them with one of my favorite ingredients, garlic scapes, which are essentially the reproductive system of the garlic plant. In fact, I like their garlicky-oniony flavor so much that I have one tattooed on my arm.

Note that the terrine must be refrigerated overnight before finishing the dish and serving.

TROTTERS

2 pork shanks
  (about 3 pounds each)

4 cups Chicken Stock
  (page 316)

1 cup dry white wine

1 cup dry Madeira

2 shallots, cut in half

2 celery ribs, cut into
  2-inch pieces

1 large carrot, cut into
  2-inch pieces

2 garlic cloves, thinly sliced

12 thyme sprigs

1 whole star anise

1 fresh bay leaf

2 tablespoons chopped
  flat-leaf parsley

Grated zest of ½ lemon
  (use a Microplane)

Kosher salt

Freshly ground black pepper

3 large eggs

1 tablespoon Husk Beer Mustard
  (page 237) or Dijon mustard

4 cups panko bread crumbs

2 tablespoons canola oil or
  Rendered Fresh Lard
  (page 316)

PEAS

2 cups fresh peas

1 cup Vegetable Stock
  (page 316)

1 tablespoon unsalted butter

1½ teaspoons kosher salt

GARLIC SCAPE PUREE

1 tablespoon unsalted butter

1 pound garlic scapes (about 6),
  washed in several changes of
  water and thinly sliced

½ cup Vegetable Stock
  (page 316)

1 tablespoon heavy cream

1 teaspoon kosher salt

PORCINI

3 large porcini mushrooms,
  (about 1 pound), washed,
  dried, halved lengthwise,
  and scored on the cut side
  in a crosshatch pattern

Kosher salt

2 tablespoons unsalted butter

1 cup Vegetable Stock
  (page 316)

Pea shoots (optional)

EQUIPMENT

A 10-by-3½-by-4½-inch
  terrine mold

FOR THE TROTTERS: Put the pork shanks into a medium pot, add the chicken stock, wine, and Madeira, and slowly bring to a simmer over medium-high heat. Skim off any scum that rises to the top. Add the shallots, celery, carrot, garlic, thyme, star anise, and bay leaf and bring back to a simmer, then reduce the heat to medium-low: you only want to get a very gentle bubble every once in a while. Cover the pot and braise the shanks until the meat pulls away from the bone, about 3 hours.

Remove the shanks and set aside. Strain the cooking liquid and pour it back into the pot. Bring to a simmer over medium heat and simmer, skimming all the fat and impurities that rise to the top, until reduced by half, about 1 hour. Remove from the heat.

Meanwhile, when the pork is cool enough to handle, pick all the meat and skin from the shanks and discard the bones. Shred the meat and tear the skin into bite-sized pieces. Transfer to a bowl, cover, and set aside.

Add 1 cup of the reduced liquid to the pulled pork to moisten it; it should be very moist. Add the parsley, lemon zest, 1½ teaspoons salt, and 1 teaspoon pepper and mix well. Pack the mixture in a 10-by-3½-by-4½-inch terrine mold. Cover and refrigerate overnight. *(Tightly covered, the terrine can be kept for up to 2 days in the refrigerator.)*

THE NEXT DAY, START THE PEAS: Put ½ cup of the peas and the stock in a blender and blend on high until very smooth, about 5 minutes. Pass through a sieve into a container, cover, and refrigerate. *(The puree can be refrigerated for up to 2 days.)*

FOR THE GARLIC SCAPE PUREE: Heat ½ tablespoon of the butter in a medium saucepan over high heat. When the butter foams, add the scapes, reduce the heat to medium-high, and cook until the scapes are tender, about 2 minutes. Add the vegetable stock and bring to a simmer.

Transfer the mixture to a blender and add the remaining ½ tablespoon of butter, the cream, and salt, and blend on high until very smooth, about 5 minutes. Return the puree to the saucepan, cover it with a lid or parchment paper, and keep it warm over low heat for no longer than 20 minutes. *(If the puree is held for too long, it will turn brown. But you can make the puree ahead, cover, and refrigerate it for up to 3 days; reheat it, stirring, over low heat, when ready to serve.)*

WHEN READY TO COOK THE TROTTERS: Unmold the terrine and cut into 6 slices.

Whisk the eggs and mustard together in a wide shallow bowl until smooth. Put the panko in another wide shallow bowl and season with salt and pepper. Dip the trotters into the egg mixture, letting the excess drip back into the bowl, then into the bread crumbs, and put on a plate.

Heat a large nonstick skillet over high heat. Add the canola oil. When the oil shimmers, carefully place the trotters in the pan and cook them until they are golden brown on the bottom, about 4 minutes. Turn them over and cook until golden brown on the other side, about 4 minutes more. The trotters are best if used at once, but you can keep them warm in a 200°F oven for up to 10 minutes.

TO FINISH THE PEAS: Heat a small saucepan over high heat. Add the butter and the remaining 1½ cups peas. Cook for 1 minute. Add the salt, reduce the heat to medium-high, add the pea puree, and gently bring to a simmer.

MEANWHILE, FOR THE PORCINI: Heat a large skillet over high heat. Season the porcini with salt. Add the butter to the skillet and when it begins to foam, add the porcini cut side down. When the porcini are golden brown, about 4 minutes, add the stock. Turn the porcini over, bring to a simmer, and cook until they are tender, about 4 minutes.

TO COMPLETE: Spoon a ¼-cup pool of the scape puree slightly off center onto each of six warm plates. Place a trotter fritter to the left of each portion of the puree and scatter some peas over the plate. Place half of a porcini in the scape puree on each plate and garnish the plates with pea shoots, if they are in season.

PORK BELLY WITH HERBED FARRO, PICKLED ELDERBERRIES, CHANTERELLES, AND SUMAC

SERVES 4

I learned a key lesson about cooking while raising pigs: pay attention to what's around you and why it's happening. When I was first developing custom feed for my pigs, I would always taste it—you should have seen the look on other farmers' faces when they saw me eating pig feed! I thought it was pretty tasty—in fact, it is almost like porridge, since it consists of corn, oats, barley, and sorghum, among other things. So I thought, "Why not serve pork belly with those same flavors?"

Here I pair the pork with things I can find growing on Wadmalaw Island: wild herbs, eucalyptus, mushrooms, and elderberries. The sumac found its way onto the plate when I was riding around Adam Musick's farm one day. He kept complaining about the wild sumac that was raising havoc on his fields, and when I told him that I cooked with it, he said, "Well, we got plenty here. Help yourself, you crazy bastard!"

Note that the pork must be brined for 24 hours, then compressed in the refrigerator overnight or 12 hours.

PORK BELLY

1 tablespoon dark brown sugar

½ teaspoon TCM (see Note and Resources, page 326)

5 tablespoons kosher salt, plus more for seasoning

One 1-liter bottle spring water

One 2-pound piece fresh pork belly (with skin)

2 tablespoons ground sumac (see Resources, page 326)

Freshly ground black pepper

1 tablespoon canola oil

Pork Jus (page 320)

HERBED FARRO

4 cups Pork Stock (page 319) or Chicken Stock (page 318)

1 tablespoon canola oil

1 cup small dice onion

1 garlic clove, minced

1½ cups Anson Mills Farro Piccolo (see Resources, page 326)

CHANTERELLES

1 tablespoon Rendered Fresh Lard (page 316)

20 small chanterelles, stems scraped clean, caps swirled in warm water to clean and dried

Kosher salt

Sherry vinegar

HERB PUREE

1 bunch chervil

1 bunch flat-leaf parsley

2 teaspoons Rendered Fresh Lard (page 316)

½ lemon

Kosher salt

1 cup Pickled Elderberries (page 223)

Ground sumac

FOR THE PORK BELLY: Combine the brown sugar, TCM, and 5 tablespoons salt in a saucepan, cover with just enough bottled water to dissolve everything, and heat, stirring, over medium heat until the sugar and salt dissolve. Remove from the heat, add the remaining water, and let the brine cool completely.

Place the pork belly in a container, pour the brine over it, cover, and refrigerate for 24 hours.

THE NEXT DAY: Preheat the oven to 200°F.

Rinse the pork belly under cold running water for 10 minutes. Dry completely, then season with the sumac and a little salt and pepper. Place the belly skin side up in a 9-by-11-inch or similar size baking dish and cook it, uncovered, for 8 hours. No basting is necessary—the amount of fat in the belly makes it self-basting.

Transfer the pork to a rimmed baking sheet and cool to room temperature. Pour the fat and juices from the baking dish into a container and reserve in the refrigerator.

Wrap the cooled pork securely in plastic wrap. To improve its texture, compress the pork by placing several heavy baking sheets on top of it to weight it down. Refrigerate it, with the weight, for 12 hours.

Divide the pork into 8 pieces. Cover and refrigerate for up to 5 days.

FOR THE HERBED FARRO: Heat the stock in a medium saucepan until hot. Partially cover, and keep warm.

Heat the oil in a medium saucepan over medium-high heat. When the oil is shimmering, add the onion and cook, stirring occasionally, until translucent, 5 to 7 minutes. Add the garlic and cook, stirring, until soft, about 1 minute. Add the farro, stirring to coat the grains with the oil. Add ¼ cup of the stock, reduce the heat to medium, and stir the farro until the liquid is almost absorbed. Continue cooking and adding ¼ cup of stock at a time, stirring frequently to prevent scorching and letting each addition be absorbed before adding the next, until the farro grains have expanded and are al dente, 45 minutes to 1 hour. The farro will look creamy, like risotto. Cover the saucepan and keep the farro warm on the back of the stove. (The farro can be made up to 3 days ahead, cooled, covered, and refrigerated; reheat, stirring, over low heat.)

FOR THE CHANTERELLES: Heat the lard in a large skillet over high heat. Add the chanterelles and cook, tossing, until soft, about 2 minutes. Season with salt and a splash of sherry vinegar. Keep warm on the back of the stove.

CONTINUED

FOR THE HERB PUREE: Pick all the leaves from the chervil and parsley stems. Reserve 24 stems of each in ice water to garnish the dish.

Bring a large saucepan of salted water to a boil. Make an ice bath in a bowl with equal parts ice and water. Put the herb leaves in a large strainer and submerge them in the boiling water for 40 seconds, then remove and submerge them in the ice bath until completely cold. Remove from the ice bath, shake off the excess water, then drain and dry on paper towels.

Blend the herb leaves in a blender on high until smooth, adding just enough water to make the blade spin. Strain through a fine-mesh sieve.

FINISH THE PORK: Heat the oil in a large skillet over high heat until shimmering. Add the pieces of pork belly (work in batches if necessary) and cook, turning once, until golden brown, about 2 minutes on each side. Drain on paper towels.

While the pork belly is cooking, add the herb puree to the farro and heat over low heat. Warm the pork jus in a small saucepan over medium heat. Reheat the chanterelles if necessary.

Shake the excess water off the parsley and chervil stems, pat dry, and put in a bowl. Warm the lard and toss the stems with a little of the lard, a spritz of lemon juice, and a little kosher salt.

TO COMPLETE: Spoon some farro onto one side of each plate and put 2 pieces of pork belly next to it. Sauce the pork belly with the pork jus. Divide the chanterelles and elderberries among the plates. Sprinkle the plates with sumac and garnish with the chervil and parsley stems.

=== **NOTE** ===

TCM, or pink curing salt, is a preservative that keeps meat from oxidizing. It's 94 percent salt and 6 percent sodium nitrite. The nitrite also helps prevent the growth of the botulinum toxin. It's important to follow the amounts specified precisely; an accidental overdose of TCM would not be good for you. In fact, it's colored pink so it won't be mistaken for table salt and sprinkled over your French fries.

CRAIG ROGERS

One day the phone rang in the chef's office at McCrady's. It was a little late in the evening, the time when most of the annoying phone calls come with people trying to sell us stuff that we would never serve. This call was different.

On the other end of the line was Craig Rogers, and as he introduced himself, I heard passion and enthusiasm in his voice. He told me all about his lamb and his farm, about the types of grass, and the breeds of lamb. Sure, this is what most purveyors do, but Craig farms in my home state of Virginia, so I paid closer attention. I told him about a chicken project that I was starting with a local farmer and my excitement about the possibility of producing heirloom breeds of chicken and serving them in the restaurant. Craig gave me a thirty-minute earful about what's wrong with the poultry industry these days and how it can be fixed. And I learned that the breed I was so excited about raising isn't even a true heirloom breed. I could tell Craig had moxie, and so I agreed to have some of his samples delivered. He sent me both lamb and heritage poultry.

Now, if you've never been in a kitchen when an ingredient changes the game, then let me tell you, it's something to behold. I love it when I gather the kitchen team around to dive into something new and the whole room goes silent. There's just an edge to it, a feeling that now so many things will have to be reconsidered and so many possibilities exist. It's simply incredible to watch people taste real food for the first time, when their eyes widen and the light turns on. It's even more incredible to see the same expression cross the faces of experienced chefs.

That's the story of Craig's lamb. It changed my kitchen, and we haven't used lamb from any other farm since. In fact, I'd never dream of it. The reason is simple: Craig Rogers produces the best lamb in America.

The value of Craig's work comes from doing things in a tried-and-true fashion. Craig working a flock of sheep with his dog is fascinating to watch. He controls a flowing crowd of sheep with just the dog and a whistle, signaling commands with alacrity. There are no electric prods or stressed animals, because Craig is a true shepherd.

Every year, I cook at dozens of charity events alongside many talented chefs. It's always interesting to listen to the chatter in the kitchen when several chefs get together. It used to be that people talked about various breeds of pork, or the latest oven equipment. Professional cooking is always a show of one-upmanship. But you never heard much about lamb before Craig. These days, if you bring another type of lamb to these events, people just shake their heads, right after you hear the phrase: "Is that Craig's lamb?" And that is all the explanation that is often needed.

SLOW-COOKED LAMB RACK WITH SPRING FAVAS, MALTED BARLEY, AND ROASTED CHANTERELLES

SERVES 4

The secret to a perfect roast is temperature control. Use a digital probe thermometer that beeps when the meat reaches a specific temperature—the probe stays in the meat while it's cooking. Most of these have an alarm setting that will tell you when your meat is at exactly the temperature you want it.

Of course, we don't cook this way in a professional kitchen. There it's all feel, sound, and smell to know when meat is cooked, but it takes years of experience to gain this sort of skill. When you remove the meat you've cooked from the oven, give it a feel: push on it a little and pay attention to the resistance. It's important to feel both ends as well as the middle of the cut. As the meat cooks, its texture will firm up, moving from a mushy raw to, if you take it that far, a firm, rubbery well-done. Of course, using a thermometer is a great way to teach yourself what your favorite meat temperatures feel like and ensure that you don't ruin the meat.

NOTE

When cooking the chanterelles, it is important to realize that every batch of chanterelles will have a different moisture content. It is critical to make sure that the liquid they release has completely evaporated before adding any other liquid.

MALTED BARLEY CRISPS

1½ cups all-purpose flour

1 tablespoon plus 1 teaspoon black malted barley powder (see Resources, page 326)

1 tablespoon confectioners' sugar

1 teaspoon kosher salt

3½ sticks (14 ounces) unsalted butter, very soft

1 tablespoon sorghum (see Resources, page 326)

LAMB

One 8-bone rack of lamb, bones cleaned, frenched (you can have the butcher do this), and wrapped in foil

Kosher salt and freshly ground black pepper

Canola oil

10 thyme sprigs

10 rosemary sprigs

5 garlic cloves, lightly smashed, skin left on

CHANTERELLES

3 tablespoons unsalted butter

20 medium chanterelles, stems scraped clean, caps swirled in warm water to clean, dried, and cut in half

2 thyme sprigs

About ¾ cup Vegetable Stock (page 316)

FAVA BEANS

1½ cups Vegetable Stock (page 316)

2 cups peeled fresh fava beans

Kosher salt

1 tablespoon chopped tarragon

1 tablespoon unsalted butter

4 tablespoons Basic Meat Sauce (page 321), warmed

FOR THE MALTED BARLEY CRISPS: Preheat the oven to 400°F.

Combine the flour, malted barley powder, confectioners' sugar, and salt in a food processor and process to combine. Gradually add the butter and then the sorghum, pulsing to combine. The dough should be very smooth.

Using a rolling pin, roll the dough out between two sheets of parchment paper to an ⅛-inch thickness. Remove the top piece of paper. Slide the bottom piece with the dough onto a baking sheet. Bake the malted barley for 10 minutes, or until dry and crispy. Cool on a rack.

Cut or break it into 1-inch squares. (The pieces will keep in an airtight container for up to 2 days.)

FOR THE LAMB: Preheat the oven to 250°F.

Liberally season the lamb with salt and pepper. Heat a 12-inch cast-iron skillet over high heat. When the skillet is hot, add ¼ inch of canola oil. When the oil begins to smoke, add one rack fat side down and sear until golden brown on the first side, 3 to 5 minutes. Repeat on all sides. Remove from the skillet and sear the second rack.

Put a wire rack in a roasting pan and cover it with the thyme, rosemary, and garlic. Place the lamb fat side up on the herbs and garlic. Put the pan in the oven and slow-roast the lamb for 45 to 55 minutes, until the internal temperature reaches 125°F for rare, or for 1 to 1½ hours, until the internal temperature reaches 135°F for medium. Remove the pan from the oven and allow the lamb to rest for 20 minutes before carving it; baste the lamb with the pan juices several times while it is resting.

MEANWHILE, FOR THE CHANTERELLES: Heat a skillet large enough to hold the chanterelles in one layer over very high heat (or work in batches if necessary). When the skillet is very hot, add the butter. As soon as the butter melts, add the mushrooms and thyme and sear the mushrooms, without moving them or shaking the pan, until all the liquid they give off has evaporated and the butter has browned, about 7 minutes. Stir in just enough vegetable stock to create a glaze, about ½ cup. The mushrooms can be kept, covered, over the lowest possible heat for up to 10 minutes.

FOR THE FAVA BEANS: Pour the stock into a medium skillet and bring to a simmer over high heat. Add the fava beans and cook until tender, about 5 minutes. Season with salt. Remove the skillet from the heat and stir in the tarragon and butter.

TO COMPLETE: Carve the lamb, slicing it between every 2 bones. Place 1 double chop on each of four warm plates. Divide the chanterelles and fava beans among the plates. Garnish with the barley crisps and meat sauce.

BRAISED LAMB NECK WITH TOMATO CONSERVE AND SQUASH SEED RISOTTO

SERVES 4

One of the tastiest parts of a lamb is most certainly the neck, full of delicious fat, and when it is properly cooked, the meat just falls off the bone. I love the ratio of fat to meat in a lamb neck. Here it's served with some characteristic flavors of summer.

Order the lamb necks from your butcher ahead of time, and note that the braised lamb must be refrigerated overnight.

LAMB NECK

2 medium lamb necks (about 2½ pounds total)

Kosher salt and freshly cracked black pepper

1 cup all-purpose flour

Canola oil

One 1-liter bottle Cabernet Sauvignon

4 cups Chicken Stock (page 318)

2 cups large dice sweet onions

1½ cups ½-inch-thick carrot slices

4 celery ribs, leaves discarded and cut into 1-inch pieces

6 garlic cloves, smashed and peeled

8 thyme sprigs

1 fresh bay leaf

1 orange, cut into quarters

1 Bosc pear, cut into quarters and cored

Squash Seed Risotto (page 57)

1 cup Heirloom Tomato Conserve (page 223)

FOR THE LAMB: Preheat the oven to 350°F.

Season the lamb necks with salt and pepper and dredge them in the flour. Heat a large heavy roasting pan over high heat on the stovetop. Add ¼ inch of canola oil. When the oil shimmers, add the necks and cook for 1 minute, then reduce the heat to medium-high and brown the necks on all sides. Remove them from the pan and set aside.

Discard the canola oil in the roasting pan. Add the wine, increase the heat to high, and scrape the bottom of the pan to dislodge the browned bits. Bring the wine to a boil and cook until it is reduced by half, about 10 minutes.

Reduce the heat to medium-high, add the chicken stock, bring to a simmer, and cook until it is reduced by one-quarter, about 10 minutes.

Remove the roasting pan from the heat and put a flat rack in the pan (with the wine and chicken stock still in it). Put the lamb necks on the rack, cover the pan, and place it in the oven. Cook the lamb for 1½ hours.

Uncover. Add the onions, carrots, celery, garlic, thyme, bay leaf, orange, and pear. Place the uncovered pan back in the oven and cook the necks until the meat pulls away easily when a knife is inserted next to the bone, about 1½ hours.

Carefully transfer the lamb necks to a baking pan and set aside. Strain the braising liquid into a bowl. Allow the liquid to cool on the countertop, skimming the fat as it rises to the top; this will take a couple of hours.

Meanwhile, when the lamb necks are cool enough to handle, carefully remove the meat from the bones, keeping it as intact as possible; discard the bones. Put the meat in a container and pour any juices from the baking pan over it.

When all of the fat has been removed from the braising liquid, pour it over the lamb. Cover and refrigerate overnight. The lamb will soak up some of the liquid and remain moist.

THE NEXT DAY, remove the lamb from the bowl and divide it into 4 pieces, keeping them as intact as possible. Put the lamb on a plate and set aside at room temperature.

Set a large skillet over high heat and pour the braising liquid from the lamb into it. Bring to a light simmer, then reduce the heat to medium-high and cook, skimming the fat as it rises, until the liquid reduces enough to coat the back of a spoon, about 20 minutes.

Strain the liquid and return to the pan. Add the lamb and bring the liquid to a light simmer, basting the meat with it as it heats.

TO COMPLETE: Divide the lamb and risotto among four warm plates. Sauce the lamb with some of the braising liquid and top it with conserve.

- But don't overcook the meat. There is a very special moment in barbequing where each protein gives up, submits, and falls to pieces but is still full of fat and moisture. This is one of the reasons it's important to keep a notebook. Your smoker may act differently from everyone else's. It's important to get to know your smoker. Treat it nice, and it'll return the favor.

- Let the cooked meat rest out of the smoker for at least 10 minutes before tearing into it.

- Have plenty of cold beer on hand and plenty of long half-true stories to tell around the smoker. Music is essential too. Before you start cooking, make a smoking playlist. My friend Rodney Scott has the most amazing barbeque playlist in existence. The music makes you feel good and the tasty barbeque follows suit.

- Have fun! If you aren't having fun cooking with smoke, something is wrong.

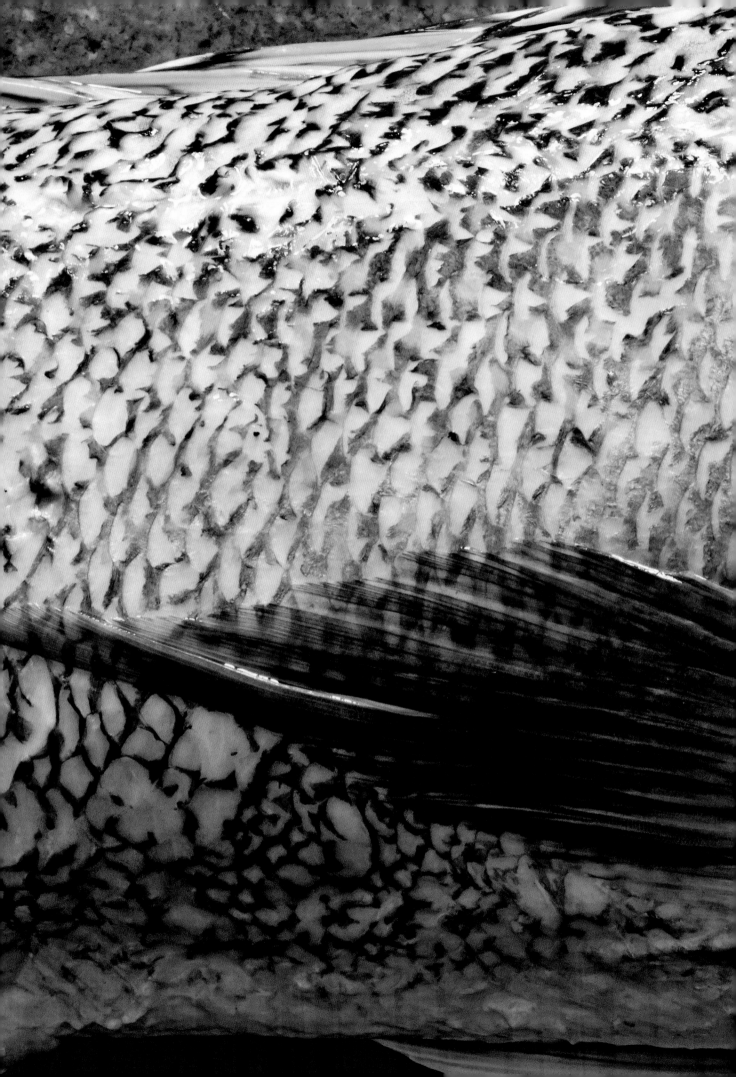

THE CREEK AND THE SEA

LOWCOUNTRY SEAFOOD

Most people want their seafood just like they want their vegetables: fresh and locally sourced. One might think that a great port city like Charleston would always have enjoyed great fresh seafood, that the harbor would have provided clams and oysters, shrimp, and lots and lots of fish. For various reasons, however, this has not always been the case. Even back before the Civil War, David Shields has told me, the outlay of commercial seafood in Charleston left a lot to be desired. One thing is for sure, though. People in the Lowcountry have always gone down to the creek to fish.

When I first came to Charleston, I discovered the sheepshead. It's an elusive fish best caught by hooking a fiddler crab as a lure and tossing it in around rock piles and docks, places where the sheepshead like to peck off barnacles with their odd human-like teeth. On many afternoons, I would steal down to the creek to cast a line and try my hand at catching sheepshead. It's a difficult task, because you can barely feel them bite. And they can be infuriating to chase around, but there's something magical about the Lowcountry that becomes clear and present while sitting on a creek bank in the marshes. There you are just the latest in a long line of folks who have come to the creek to try and catch that night's dinner, and you learn how to do so by watching those who have done so for years. Because of restoration efforts, there is now an abundant supply of shrimp and soft-shell crabs in our area, but modern tastes, ruined by a love of fish sticks and commercial tartar sauce, don't stoke much demand for unusual seafood. So these days, we're just beginning to tap the diversity that lies beyond the harbor and in the marshes. Charlestonians are again experimenting with seafood: periwinkles from the marsh, local wreckfish and rudderfish from deep waters, and many other varieties that traditionally were discarded as trash fish.

Today there are people like Sam Ray around. He fishes the "Charleston Bump" for wreckfish (see page 166). It's a deep-water mountain about seventy miles off the coast of Charleston, so large and jagged that it diverts the entire Gulf Stream. Here, clinging to the protection of rocky crags, are those delicious fish. Sam spends days at sea fishing in a thousand feet or more of water, pinpointing fish that must be precisely targeted. If the rig is off by thirty yards, they won't bite, and then it takes an hour to relocate the boat. The payoff is a delicious morsel, sustainably harvested, and very specific to the Lowcountry, since it's almost the only place in the America where you'll find wreckfish coming ashore.

Finding great seafood in the Lowcountry means focusing on people and relationships. No one, whether a chef or a retail customer, is going to find interesting products without searching for those who are experts in the field. If you live in a place with a wonderfully diverse fresh market, you may only need to become friends with the best fishmonger available. But if you have the opportunity, take yourself to the water's edge, talk to an old salt, and seek out the day's fish haul. That's how you'll discover some great fish.

NOTABLE SEAFOOD OF THE SOUTHERN WATERS

Although fish wasn't always abundant in the waterways of Charleston, nowadays we're lucky enough to be able to feast on a great variety of seafood.

WRECKFISH

There are only a couple of people on the East Coast who harvest wreckfish commercially. So it's not easy to get the fish outside of the Charleston local market—but if you find it, you'll have one of the tastiest fish of the Lowcountry. Most people compare wreckfish to grouper, but it has a more delicate flavor, and a pleasantly firm texture, shaped by the cold depths of the Charleston Bump where it lives, more than a thousand feet beneath the Atlantic.

TRIGGERFISH

Triggerfish like to steal bait from fishermen, their skin is rough and leathery, and the fillets are small. But they are delicious, with a subtle white flesh that takes well to a variety of preparations. Because triggerfish was considered undesirable for many years, it used to be cheap, but not anymore. Nowadays chefs want triggerfish the way they used to covet red drum or orange roughy, and the price reflects its popularity across the Eastern Seaboard.

AMBERJACK

The amberjack has a nasty reputation. If you ever hook one on a rod and reel, you'll see its propensity for violence and aggression. And if you land a good-sized one and take it home to eat, you'll see why most people won't touch amberjack. These suckers are full of worms. But although the amberjack in Charleston's warm waters may harbor parasites, the squeamish will miss out on a delicious treat. Soaking the flesh in a little salt brine overnight will drive out the worms, and this strong-flavored oily fish might be the best of any for the summer grill.

BLACK SEA BASS

The "black fish" was the more common fish at the Charleston market for much of history. It's flaky, sweet, and firm—and it's still the most frequent catch at the Charleston jetties. The man-made rock formations installed at the entrance to the Charleston harbor in the late 1800s were designed to divert the flow of the tides to scour the channel, clearing shipwreck debris and sediment from the bottom. The jetties did the job, while making for one of the best habitats for black sea bass on the East Coast.

SHEEPSHEAD

There is no fish trickier to catch than the sheepshead. To haul them in, first you have to catch fiddler crabs, which means laying out an old bedsheet on the mudflats surrounding the marsh and then scooping up the crabs once they've crawled atop it. Then you want to find some old dock posts or bridge structure covered with barnacles, because sheepshead like to gnaw on them. Sheepshead have teeth that look like a mammal's, with big "molars" to grind up the barnacles. It's kind of scary when you see a sheepshead's teeth, but they're gentle feeders. And this gentle giant, which swims around feeding on crab and oysters all day, ends up being a real treat, tasting much like the buttery crustaceans that it consumes.

SHRIMP

Shrimp have always been plentiful in Charleston, and, of course, they are part of the iconic shrimp and grits. The shrimp come in too many types to describe here, but some of the best ones are caught by the oldest method, free-casting a net along a muddy creek bank. Modern recreational fishermen bait larger shrimp on inland waterways with fish meal, but a shrimp that's been munching on stinky meal for several hours before you scoop him up is not your culinary friend. Creek shrimp, on the other hand, are small and sweet. Going down to the creek for hours was often the only way to get breakfast when you had no money, and creek shrimp are still a favorite poor-man's staple—they were the original shrimp in shrimp and grits.

Now Charleston has trawlers, the big picturesque shrimp boats that dot the coast. They go out into deeper waters offshore and haul their nets across the bottom of the

ocean. They catch a variety of shrimp, depending on season and conditions, but in the spring, the roe shrimp come in. They're jumbo and plump, as sweet as the small varieties from the creek, and packed with roe, or eggs. Cook these in their shell over low heat, and serve with salt and lemon. When they are simply prepared like this, you can really taste where the shrimp came from.

SOFT-SHELL CRAB

When you talk about crabs, most people think of the Chesapeake Bay. But, like the Tidewater, Charleston has dedicated purveyors who give us crab in the highest form: the soft-shell. We also support a serious trade in stone crab claws (by law, you're not allowed to keep the crab, just one claw, which will grow back) and hard-shell crabs.

When spring arrives, everyone anxiously awaits the first of the soft-shells, crabs that are entering the process of molting and come to the plate with a soft, edible exoskeleton. They're so delicious that they don't take much in the way of cooking at all. Dredge them in flour and panfry them gently in butter. They emerge salty and rich, with a deep flavor of the sea that no other crab can match. They're eaten whole, legs, body, and claws, with only the gills removed, and they go fast, because the delicious nature of a spring soft-shell is no secret in the Lowcountry.

SEA URCHIN

While most of the seafood I've described has been caught and eaten out of our waters for ages, sea urchins have long been overlooked. The Japanese are enamored with urchin roe. Why? Because the taste is surreal—a briny, unctuous morsel with the fleeting texture of panna cotta. There are sea urchins in the waters surrounding Charleston. So let this serve notice: the Charleston sea urchin is worth seeking out.

CHERRYSTONE AND LITTLENECK CLAMS

Clams come in all shapes and sizes, but the most tender and succulent are cherrystones and littlenecks. Littlenecks tend to be a bit smaller than cherrystones, which are about 2½ to 3 inches across. This is simply a function of when they are harvested and the best clammers will harvest them young. A small clam with no grit tastes perfectly of the brininess of the open sea.

BROWN OYSTER STEW
WITH BENNE

SERVES 6

This is one of those iconic Lowcountry dishes that fell by the wayside when people stopped growing benne in the twentieth century. The recipe simply didn't work the same way or taste good with modern sesame seeds. So this version exemplifies cooking with what you have: it makes use of the by-product of pressing benne seeds for oil. The resulting mash is dried and milled as a flour. It thickens the soup and adds a wonderful nuttiness. This is a perfect bowl of food for a cold winter day. Serve it over Carolina Gold rice or grits.

40 Eastern oysters, preferably Caper's Blades (see Resources, page 326)

1 cup diced bacon, preferably Benton's (see Resources, page 326)

1 cup medium dice white onion

½ cup Anson Mills Antebellum Bennecake Flour (see Resources, page 326)

½ cup Anson Mills French Mediterranean White Bread Flour (see Resources, page 326)

3½ cups Chicken Stock (page 318)

Juice of 1½ lemons (about ⅓ cup)

Kosher salt

Husk Hot Sauce (page 238) to taste

2 tablespoons Anson Mills Antebellum Benne Seeds (see Resources, page 326)

½ cup finely sliced scallions

Shuck the oysters, being careful to save all their liquor. Place the oysters and liquor in a container, cover, and refrigerate.

Put the bacon in a large heavy-bottomed saucepan and cook over medium heat, stirring frequently, until the fat is rendered and the pieces of bacon are crispy, about 8 minutes. Add the onion and cook, stirring occasionally, until translucent, 5 to 7 minutes. Reduce the heat to low, add both flours, and, using a wooden spoon, stir until the flours absorb the bacon fat. Then stir the roux constantly for about 7 minutes, until it is a dark golden color.

Drain the oysters, reserving the liquor. Add the liquor and chicken stock to the roux, whisking to combine, and, still over low heat, bring to a simmer. Simmer until thickened, about 10 minutes.

Add the lemon juice and season with salt and hot sauce. Add the oysters and cook, without stirring, until just warmed through, about 3 minutes.

Garnish with the benne seeds and sliced scallions.

STONE CRAB COCKTAIL
WITH SMOKED MAYONNAISE, AVOCADO, AND LOVAGE

**SERVES 6
AS AN APPETIZER**

My friend and crab supplier Kimberly Carroll is really passionate about crabs. She wears T-shirts with crabs on them, gold crab necklaces, and even crab earrings! When she calls the kitchen after a day on the water, you can hear the enthusiasm in her voice: "Hey there, Sean! Got a pile of stoneys today! How many ya want?"

Kimberly's crabs arrive in the kitchen straight off the boat, still very much alive and pinching. This dish is a simple riff on a crab cocktail with a smoked mayo that goes great with the crab and a bit of avocado and lovage. When you present food so simply, the quality really matters, so only make this with the freshest crabs.

SMOKED OIL
2 cups grapeseed oil

SMOKED MAYONNAISE
2 large eggs

1½ teaspoons fresh lemon juice

1 teaspoon Husk Beer Mustard (page 237) or Dijon mustard

1½ teaspoons lemon oil

5 tablespoons Smoked Oil (above)

Kosher salt

SALAD
8 large stone crab claws (about 5 pounds), cracked, meat removed from the shells, and picked clean

2 ripe avocados

Juice of 1 lime

1 cup thinly sliced lovage leaves

1 tablespoon extra-virgin olive oil

EQUIPMENT
Charcoal grill

3 pounds hickory wood chips, or as needed, soaked in cold water for a minimum of 30 minutes but preferably overnight

FOR THE SMOKED OIL: Prepare a hot fire in a charcoal grill.

Add 2 cups soaked wood chips to the hot coals. Put the grapeseed oil in a medium saucepan and place the saucepan on the grill grate. When the chips begin to smoke, close the lid and let the oil smoke until it has a distinct smoky taste, about 2 hours. Check every 30 minutes, and if the smoke has dissipated, add more soaked chips (see sidebar, page 160). Remove the saucepan from the grill and let the oil cool to room temperature.

Pour the oil into a container. *(Tightly covered, the oil will keep for up to 2 months in the refrigerator.)*

FOR THE MAYONNAISE: Bring 4 cups water to a boil in a medium saucepan over high heat. Meanwhile, make an ice bath in a bowl with equal parts ice and water. Add the eggs to the boiling water and cook for 5 minutes. Drain the eggs and plunge them into the ice bath to cool. Carefully peel the eggs under running water; the yolks should still be fluid. Pat dry.

Put the eggs, lemon juice, and mustard in a blender and blend on medium until smooth. Slowly add the lemon oil and the 5 tablespoons smoked oil, blending to emulsify into a mayonnaise. Season with salt. Transfer to a container, cover, and refrigerate until chilled. *(Tightly covered, the mayonnaise will keep for up to 3 days in the refrigerator.)*

TO COMPLETE: Dress the crab with ¼ cup of the smoked mayonnaise.

Peel, halve, and pit the avocados. Cut them into ¼-inch dice and toss them with the lime juice in a small bowl. Add the lovage leaves and olive oil and gently stir to combine. Divide the avocado and lovage mixture among six small bowls, arrange the crab on top, and serve.

NOTE

This recipe makes more smoked oil than you will need for the mayonnaise, but the oil keeps well in the refrigerator and can be used in vinaigrettes or to poach fish.

PICKLED SHRIMP
WITH CILANTRO AND FENNEL

SERVES 6
AS AN APPETIZER

Pickled shrimp is a Charleston classic, and at many a cocktail party, you will find a jar of shrimp lolling in sharp, acidic brine. This is not unlike South American ceviche, except ours usually gets its acidity from lemon and vinegar, or both, instead of lime. In this version, I punch up the intensity with the seeds of the cilantro plant and the anise-like pungency of fennel, which always goes well with fish or shellfish. Even though the shrimp will keep for a week in the refrigerator, they're best eaten straight-away, before the acid begins to break down their texture. Of course, if you serve these to guests, there will not be any left for tomorrow.

Although the recipe is fairly simple, you will need to make the pickling liquid 2 days ahead and then let the shrimp pickle overnight.

PICKLING LIQUID

1 large Vidalia onion, shaved as thin as possible

1 small fennel bulb (about 12 ounces), trimmed fronds reserved for garnish, and bulb shaved as thin as possible

1 celery rib, shaved as thin as possible

1 small carrot, shaved as thin as possible

4 garlic cloves, shaved as thin as possible

1 jalapeño, seeded and shaved as thin as possible

1 cup apple cider vinegar

Grated zest of 1 lemon (use a Microplane)

Grated zest of 1 lime (use a Microplane)

¾ cup fresh lemon juice

¾ cup fresh lime juice

½ cup fresh orange juice

2 tablespoons extra-virgin olive oil

2 tablespoons celery seeds

2 tablespoons yellow mustard seeds

2 tablespoons fennel pollen (see Resources, page 326)

1 tablespoon fennel seeds

1 teaspoon black peppercorns

1 teaspoon turmeric

1 teaspoon crushed red pepper flakes

3 fresh bay leaves

SHRIMP

2 quarts Vegetable Stock (page 316)

2 cups dry white wine

2 tablespoons kosher salt

1 tablespoon crushed red pepper flakes

2 fresh bay leaves

2 pounds large shrimp (16–20 count), peeled and deveined

Fennel flowers and fronds (optional)

Cilantro blossoms, berries, and/or leaves (optional)

Carrot flowers (optional)

FOR THE PICKLING LIQUID: Combine all of the ingredients in a medium bowl. Cover and refrigerate overnight so that the flavors can meld.

FOR THE SHRIMP: Combine the stock and wine in a large pot and bring to a simmer over high heat. Add the salt, red pepper flakes, and bay leaves, reduce the heat to medium, and simmer for 20 minutes.

Add the shrimp to the poaching liquid and cook for 2 minutes, until they turn pink. Drain the shrimp, reserving the broth, if desired. *(You can freeze the broth for up to 3 months to use again for poaching seafood.)*

Drop the shrimp into the cold pickling liquid. The citrus in the pickling liquid will continue to "cook" the shrimp. Cover and refrigerate the shrimp overnight. *(Tightly covered, the shrimp will keep for up to 1 week in the refrigerator.)*

Serve the shrimp in a large bowl (drain the liquid out first) or in individual Mason jars. Garnish with the reserved fennel flowers and fronds; cilantro blossoms, berries, and/or leaves; and carrot flowers, if you have them.

STONE CRAB
WITH CUCUMBER JUICE, FENNEL JELLY, AND RAW APPLE

SERVES 6
AS AN APPETIZER

The stone crab is a voracious-looking fellow. It's brown, and one of its claws is enormous. A large stone crab can take off a finger right at the knuckle. Stone crabs are also relatively rare, and regulations require that the crab be thrown back when caught, after the fisherman removes the large claw for the catch. The crabs regrow the lost claws, and we get the very succulent meat. Stone crab meat has a decidedly clean flavor, almost mineral in character, and so I like to serve it simply, with just a few other flavors to complement it.

Note that the fennel consommé must drain for about 48 hours before you can finish the fennel jelly.

=== NOTE ===

You can freeze the extra jelly and use it as a broth for seafood.

FENNEL JELLY

2 large fennel bulbs (about 2 pounds), trimmed and diced; reserve a few fronds for garnish

3 sheets silver-strength gelatin (see Resources, page 326)

Juice of ½ lemon

¾ teaspoon agar-agar powder (see Resources, page 326)

CUCUMBER JUICE

1 large English cucumber, peeled

¼ teaspoon xanthan gum (see Resources, page 326)

¼ teaspoon kosher salt

STONE CRAB

6 large stone crab claws (about 4 pounds), cracked, meat removed from the shells and picked clean

Juice of ½ lemon

1 tablespoon extra-virgin olive oil

2 teaspoons kosher salt

½ teaspoon piment d'Espelette (see Resources, page 326)

RAW APPLE

1 Granny Smith apple

1 tablespoon fresh lemon juice

Kosher salt

Olive oil

Piment d'Espelette (see Resources, page 326)

EQUIPMENT

Juice extractor

Butcher's twine

Two 12-by-18-inch rimmed baking sheets

3-inch round cookie cutter

FOR THE FENNEL CONSOMMÉ: Run the fennel bulbs through a juice extractor. Strain the juice through a fine-mesh strainer; reserve 2 cups juice.

Place 1 gelatin sheet and ½ cup cold water in a bowl and let the gelatin soften for 2 minutes. Warm 1 cup of the fennel juice in a saucepan. Lift the gelatin from the water and gently wring it out, then add it to the warm fennel juice and heat over medium-low heat, stirring, until dissolved, about 1 minute. Add to the remaining fennel juice, along with the lemon juice, and pour into a baking pan. Place it in the freezer until solid.

Fit a strainer over a bowl and line it with a double layer of cheesecloth. Scrape the frozen fennel juice into the cheesecloth and tie the ends together with twine to form a sack. Attach the sack to a dowel or the handle of a wooden spoon and suspend the sack over the bowl. Refrigerate and allow to drain for about 48 hours; you need 1 cup plus 2 tablespoons fennel consommé to make the fennel jelly.

FOR THE FENNEL JELLY: Spray two 12-by-18-inch rimmed baking sheets with nonstick spray, then wipe clean with a paper towel.

Place the remaining 2 gelatin sheets and 1 cup cold water in a shallow bowl and let the gelatin soften for 2 minutes. Warm ½ cup plus 2 tablespoons of the fennel consommé in a small saucepan. When the gelatin sheets are soft, lift them from the water and wring them out, then add them to the warm fennel consommé and heat over medium-low heat, stirring, until the gelatin has dissolved, about 1 minute. Remove from the heat and add the remaining ½ cup fennel consommé, then pour into a blender, turn it on to medium speed, and allow a vortex to form. Sprinkle in the agar-agar and blend for 3 minutes.

Pour the mixture back into the saucepan, bring to a boil over high heat, and boil for 2 minutes to hydrate the agar-agar. Pour enough of the mixture onto the prepared baking sheets to just cover the bottoms. Refrigerate for at least 1 hour, or up to 1 day.

FOR THE CUCUMBER JUICE: Run the cucumber through a juice extractor; you need 1 cup juice. Strain the juice through a fine-mesh sieve into a blender. Turn it on to medium speed and allow a vortex to form. Sprinkle in the xanthan gum and blend for 2 minutes. Add the salt. Pour into a container and refrigerate for at least 30 minutes, or up to 2 hours.

FOR THE STONE CRAB: Gently combine the ingredients in a bowl. Cover and refrigerate.

FOR THE APPLE: Peel and core the apple and cut it into ¼-inch dice. Toss with the lemon juice and season with salt.

TO COMPLETE: Divide the crab among six bowls. Cut the fennel jelly into six rounds using a cookie cutter. Place a round on each dish and top with cucumber juice, apple, fennel fronds, and piment d'Espelette.

"CLAMMER DAVE"
BELANGER

"Clammer Dave" cares about his clams the way we care about the plates that leave our kitchen. Each clam is polished and purged of any and all grit or sand. They're perfect, and they're produced sustainably, seeded into a mudflat that Dave leases from the state of South Carolina, babied and coddled until they're just the right size, then harvested, cleaned, and bagged for delivery.

The first time I took a clam delivery from Dave I asked him what he thought the best way to purge them was. He laughed and said that there was no need. In fact, he promised that if I ever found a dirty one, he would give me a hundred clams free. So it goes almost without saying that I fell in love with Dave's clams, and we have used them ever since.

His clams are briny and tender. I like them raw, with a splash of Texas Pete or a little squeeze of lemon. They have *terroir*, a taste of Capers Inlet, from which they come.

One day Dave showed up with some oysters and a huge smile on his face. Most oysters in Charleston grow in clusters; some clusters can have fifteen to twenty oysters on them. They are salty and delicious, much smaller than the big "selects" that come in from the Gulf Coast. Some people claim that slaves used to break the native Lowcountry clusters apart, place them back in the water, and "grow" them up into larger singles. And Dave, at least, has proven that attempting such a thing is not an impossible endeavor—but it does involve grueling labor.

Up until meeting Dave, I had never seen anyone who was crazy enough to go through the trouble of breaking oysters into "singles," growing them out, cleaning them, and purging them to remove all traces of marsh grit. Spurred by the legends of slaves, Dave does just that. He produces special racks that hold the singles and submerges them in a specially selected surf zone at Capers Inlet, which, owing to its location near a large nature preserve, has some of the cleanest water in the state. The oysters elongate and sweeten in the super-oxygenated water and naturally purge themselves of the marshland flavor. They are polished like pearlescent sculptures. He calls them "Caper's Blades."

Once I went to a secret oyster spot that I'd found to try the process myself. The tide went down far enough for me to get in and pluck some oysters from the pluff mud, so I waded in with a five-gallon bucket to chisel out the first cluster. Soon there was blood everywhere. They don't call them "blades" for nothing. The oyster shells are so sharp that you don't even realize you're cut until it's too late. I was stuck in the mud with bloody hands, not having fun anymore, and I thought of Dave doing this every day. I cleaned myself up, and spent the next hour filling up several buckets with oysters. I hosed them off and took them back to the kitchen. Then I had my team chisel the oysters and clean them up to look like Dave's. This took the *entire* day, and everyone was walking around with Band-Aids on their fingers. I can promise you that after that ordeal, my team and I handled Dave's oysters with a lot more appreciation, care, and patience.

HOW TO THROW A LOWCOUNTRY SEAFOOD BOIL

In the Lowcountry, there is a long-standing tradition of throwing outdoor parties. The landscape just begs for it. I throw a lot of parties when the weather is right, and I almost always include the classic seafood boil on my menu.

A Lowcountry boil is the name we give to both the party and the food we eat there. Frogmore stew is another name for the dish. Frogmore stew originated in the Frogmore community of St. Helena Island, Georgia. But these kinds of parties happen all over the South. And it often turns into more of an event than a party. I love an occasion that involves food to be consumed standing up, using your hands. This kind of eating brings everyone together, and the conversations had over a table lined with old newspapers and covered with piping-hot food are priceless. It's also an easy party to throw. You're making a one-pot meal that doesn't require utensils or plates; what more could you ask for? The idea of a Lowcountry boil fits the way Southerners live: laid back and easy.

Cooking a Lowcountry boil can be a pretty personal thing. Everyone has their own idea of what should be included, based on how their families cooked it. At the end of the day, you can throw in whatever you want as long as it includes shrimp, potatoes, sausage, and corn. I like to use clams and live blue crabs as well. My friend Mike Lata puts big pieces of zucchini and squash in his. Here is how it works.

EQUIPMENT

A large table

Lots of old newspapers

A regular turkey fryer setup with a basket and lid and a propane tank

A large trash can

Handi Wipes

FOOD

Seafood seasoning—I prefer Old Savannah Spice Company Crab and Shrimp Boil, but Old Bay Seasoning works too (¼ cup per gallon of water, plus additional for final seasoning)

Lemons (5 per gallon of water, plus additional for final seasoning)

Peeled garlic (10 cloves per gallon of water)

Potatoes, preferably small ones, like red skins, rinsed (5 per person)

Smoked sausage, preferably andouille, cut into 2-inch pieces (½ pound per person)

Crabs (4 per person; optional)

Shucked corn, cut in half (1½ ears per person)

Clams (12 per person; optional)

Shrimp, preferably head on, in the shell

TO SERVE

Cocktail sauce (¼ cup per person)

Butter, melted (2 tablespoons per person)

Lemon wedges (3 per person)

TO PREPARE:

Line the table with the old newspapers. Set out small bowls for the cocktail sauce, melted butter, and lemon wedges.

TO COOK:

Fill the turkey fryer halfway with water. Turn the heat on to high. Add the seafood seasoning, lemons, and garlic. Put the lid on the pot and bring to a simmer; bringing this much water to a boil takes a while. Once it's at a simmer, let it simmer for 10 minutes.

Add the potatoes, cover and simmer for 10 minutes. Add the sausage. If you are using crabs, add them now. Cover and simmer for 8 minutes. Add the corn and, if you are using them, clams. Cover and simmer for 8 minutes. Add the shrimp and cook, uncovered, for about 2 minutes, depending on the size of the shrimp.

Remove the basket from the turkey fryer and dump the contents on the table. Season the pile with lemon and additional seafood seasoning.

Dig in. Douse your serving with cocktail sauce or butter or a squeeze of lemon.

There are no forks, knives, or plates, so this makes for easy cleanup. Clear the newspaper and all the leftover junk into the trash can. Provide your guests with Handi Wipes to clean up.

ROASTED SCALLOPS
with PUMPKIN
and MATSUTAKE,
BRUSSELS SPROUTS,
and BLACK TRUFFLE

SERVES 4
AS AN APPETIZER

I love cooking scallops, and I love eating scallops, with their natural sweetness and oceanic flavors. I get the best ones from North Carolina and Virginia. When a scallop is perfectly cooked, the texture is amazing.

This dish uses scallops as a backdrop to the flavors of late fall and early winter. The tail end of fall and the beginning of winter is an exciting time to cook: it's the time of year when perfect fall ingredients like Brussels sprouts and pumpkins run smack-dab into winter's most spectacular gift, the black truffle. We get our truffles from Tom Michaels in Tennessee. Matsutake mushrooms are another very special ingredient. Some people call them pine mushrooms, because they grow in pine forests and end up tasting of them as well. Most people wouldn't think of pine flavors with seafood, but the earthiness of the Brussels sprouts tempers the matsutake, and the truffle brings everything into focus.

PUMPKIN

1 small heirloom pumpkin (about 1 pound), preferably a Buckskin

Kosher salt

3 thyme sprigs

3 flat-leaf parsley sprigs

1 fresh bay leaf

½ star anise

MUSHROOMS

2 large matsutake mushrooms (about 8 ounces)

1 cup Vegetable Stock (page 316)

¾ teaspoon kosher salt

SAUCE

1 cup heavy cream

1 cup black truffle juice (see Resources, page 326)

SCALLOPS

8 large (U-10) dry-packed sea scallops (see Note)

Kosher salt

4 tablespoons unsalted butter

2 tablespoons canola oil

6 thyme sprigs

BRUSSELS SPROUTS

1 cup Vegetable Stock (page 316)

1½ teaspoons agave nectar

5 Brussels sprouts, trimmed and separated into individual leaves

¾ teaspoon kosher salt

Shavings of fresh black truffle (optional)

EQUIPMENT

Immersion circulator

Vacuum sealer

FOR THE PUMPKIN: Preheat the oven to 400°F. Place a wire rack on a rimmed baking sheet.

Cut the top from the pumpkin and discard it. Scoop out the strings and seeds and season the interior with salt. Put the thyme, parsley, bay leaf, and star anise in the cavity. Place the pumpkin on the prepared baking sheet and roast it for 45 minutes, or until tender. Remove the pumpkin from the oven, being careful not to spill the juices, and let cool slightly.

MEANWHILE, FOR THE MUSHROOMS: Clean the mushrooms with a toothbrush and warm water to remove any dirt on the caps. Scrape the stems clean with a paring knife. Cut the mushrooms into ¼-inch-thick slices.

Pour the vegetable stock into a large skillet and bring to a simmer over high heat. Remove the skillet from the heat, add the mushrooms, and season with salt. To preserve their fragrance, poach the mushrooms in the hot liquid off the heat just until tender, 1½ to 2 minutes. You may set the mushrooms aside in their liquid for up to 1 hour; reheat over low heat before serving.

FOR THE SAUCE: Put the cream and truffle juice in a medium saucepan and bring to a simmer over high heat. Reduce the heat to medium-low and simmer slowly until the sauce coats the back of a spoon, about 10 minutes. Reduce the heat to the lowest setting, cover, and keep warm. The sauce can be held for up to 30 minutes.

TO FINISH THE PUMPKIN: When the pumpkin is cool enough to handle, remove the thyme, parsley, bay leaf, and star anise and pour the juices into a container. Scoop out the flesh and place it in a blender, along with ½ cup of the juices. Blend on high until very smooth, about 5 minutes. The pumpkin can be kept warm over the lowest possible heat for up to 20 minutes. (*Alternatively, it can be cooled to room temperature, covered, and refrigerated for up to 2 days; gently reheat in a covered saucepan over low heat before serving.*)

FOR THE SCALLOPS: Preheat the water bath in an immersion circulator to 50°C (see sidebar, page 118).

Pat the scallops as dry as possible and season with salt. Place them in a vacuum bag with 2 tablespoons of the butter and seal the bag on the highest setting. Cook the scallops at 50°C for 10 minutes.

TO SEAR THE SCALLOPS: Remove the scallops from the vacuum bag and pat them dry with paper towels, making sure both sides are as dry as possible. Heat a large cast-iron skillet over high heat and add the canola oil. When the oil shimmers, add the scallops and sear on the first side until golden brown, about 2 minutes. Reduce the heat to medium, remove the scallops from the skillet, and add the thyme sprigs. Then place the scallops seared side up on the bed of thyme, add the remaining 2 tablespoons butter to the skillet, and cook the scallops, basting them with the foaming butter, until they are just cooked through, about 2 minutes.

FOR THE BRUSSELS SPROUTS: While the scallops are cooking, put the stock and agave nectar in a large saucepan and bring to a simmer over high heat. Add the Brussels sprout leaves and simmer for 1 minute. Add the salt and stir to dissolve, then drain. The leaves will be almost raw.

TO COMPLETE: Divide the pumpkin puree, scallops, and mushrooms among four warm plates. Using an immersion blender, froth the sauce by tipping the saucepan toward you and keeping the blender in the same place, then scoop the froth from the top of the sauce, and place a little on each scallop and mushroom. Carefully scatter the Brussels sprout leaves around the plates. Garnish with black truffle shavings, if you have them.

=== **NOTE** ===

Look for dry-packed scallops (or, better yet, for scallops still alive in the shell). Other types of scallops have been pumped full of additives to keep them from spoiling and to plump up their weight. Their taste is off, and they won't sear. A dry-packed scallop should be sticky to the touch—it will adhere to your down-turned palm.

LOBSTER
with PARSNIP PUREE, LEEKS BRAISED with ORANGE, and VADOUVAN JUS

**SERVES 4
AS AN APPETIZER**

I served this dish the first time that I cooked at the James Beard House. I was terrified. Not only was I cooking in such a prestigious place, but I was only twenty-four years old! I didn't overcook the lobster, and it turned out to be one of the best nights of my life. The flavors here are just as sweet as the memories I have of that night. The floral qualities of parsnips go perfectly with the sweetness of lobster and the orange-flavored leeks. The vadouvan spice mixture—with its Indian-influenced flavors of curry, cumin, and cardamom—is an unexpected element that brings everything together.

PARSNIP PUREE

2 large parsnips (about 1 pound), peeled and shaved as thin as possible

1½ cups Vegetable Stock (page 316)

1 tablespoon cream cheese

1 tablespoon unsalted butter

1 teaspoon kosher salt

LEEKS

4 baby leeks, white and tender green parts, cut lengthwise in half, sliced about ¼ inch thick, and rinsed in several changes of water

Grated zest (use a Microplane) and juice of 1 orange

1 tablespoon agave nectar

2 teaspoons kosher salt

1 tablespoon unsalted butter

LOBSTERS

1 gallon Vegetable Stock (page 316)

2 cups dry white wine

2 tablespoons white vinegar

1 tablespoon kosher salt

2 fresh bay leaves

4 live lobsters (about 1 pound each)

GARNISH

Vadouvan Jus (page 322)

Grated zest of 1 orange (use a Microplane)

4 orange segments

Sorrel and garlic flowers (optional)

Vadouvan Spice (page 309)

FOR THE PARSNIP PUREE: Put the parsnips and stock in a medium saucepan and bring to a simmer over high heat. Reduce the heat to medium, cover, and cook the parsnips until tender, about 7 minutes.

Transfer the parsnips and stock to a blender and blend on high until very smooth, about 5 minutes. Add the cream cheese, butter, and salt and blend for 1 minute more. Return the mixture to the saucepan, cover, and keep warm on the back of the stove. (*Alternatively, put the puree in a container, cool to room temperature, cover, and refrigerate. Tightly covered, the puree will keep for up to 2 days in the refrigerator. Gently warm in a saucepan over low heat when ready to serve.*)

FOR THE LEEKS: Combine all of the ingredients except the butter in a large saucepan and bring to a simmer over high heat. Reduce the heat to low, cover, and cook the leeks until they are tender, about 10 minutes. Stir in the butter. The leeks are best served right away, but they can be held over low heat for about 10 minutes.

FOR THE LOBSTERS: Combine all of the ingredients except the lobsters in a large pot and bring to a simmer over high heat. Reduce the heat to medium and simmer for 20 minutes.

Add the lobsters to the pot and set a timer for 5 minutes. After 5 minutes, remove the lobsters from the pot but keep the pot over the heat.

When the lobsters are cool enough to handle, remove the tails from the heads and claws by twisting and pulling them off. Place the lobster tails back in the pot (they need to cook longer than the claws) and cook for 2 minutes. Remove the lobster tails from the pot.

When they are cool enough to handle, crack the shells of the tails and remove the meat from the shells, keeping it in one piece. Devein the tails. Using a lobster cracker or nutcracker and kitchen shears, remove the meat from the claws and knuckles. (*In a tightly covered container, the lobster meat can be refrigerated for up to 2 days.*)

TO COMPLETE: In a small saucepan, bring the vadouvan jus to a simmer over low heat.

Place a lobster tail slightly off center on each of four warm plates. Divide the remaining lobster meat among the plates. Place a few dollops of parsnip puree around the lobster on each plate. Spoon the jus into the center of the plates. Scatter the leeks around the plates and garnish the plates with the orange zest and segments and with sorrel and garlic flowers, if you have them. Sprinkle a little vadouvan on each plate and serve.

BRAISED AND GRILLED OCTOPUS
WITH CORIANDER, EGG YOLK, FENNEL, AND VARIEGATED LEMON

SERVES 6

If I go to a restaurant and they have octopus on the menu, I always order it—especially if it's charred after being braised into submission. Octopus can be tricky to cook. Each one seems to become tender at its own rate. I cut a small piece off as I'm braising and test it. If it's not nice and tender, I simply cook it a little longer. When you're grilling it, make sure that your grill is sizzling hot so that you can get that wonderful charred flavor. This dish stays nice and light with just a little fennel, coriander, and a spritz of lemon. Charleston coriander is an herb, also called rau ram. It brings a great cilantro-like flavor that pairs well with the variegated lemon, which is less acidic than standard lemons. You can substitute regular cilantro and Meyer lemon juice in a pinch.

OCTOPUS
1 octopus (about 5 pounds)

About 2 quarts extra-virgin olive oil

1 large fennel bulb (about 1 pound), bulb and stalks cut into large dice

15 thyme sprigs

2 lemons, cut in half

2 tablespoons toasted coriander seeds

1 teaspoon kosher salt, plus more for seasoning

EGG YOLKS
Yolks from 12 large eggs

Kosher salt

Extra-virgin olive oil

PARSNIPS
2 cups Parsnip Puree (page 182)

2 tablespoons squid ink (see Resources, page 326)

Kosher salt and freshly ground white pepper

FENNEL
1 large fennel bulb with stalks (about 1 pound), trimmed; reserve some fronds for garnish

2 tablespoons extra-virgin olive oil

3 drops fennel extract (see Resources, page 326)

Kosher salt

1 cup Charleston coriander or cilantro leaves, plus more for garnish

Juice of 1 variegated lemon or Meyer lemon

Fennel pollen, fronds, and flowers (see Resources, page 326)

EQUIPMENT
Immersion circulator

Vacuum sealer

FOR THE OCTOPUS: Preheat the oven to 200°F.

Place the octopus in a deep roasting pan and cover it with the olive oil. Scatter the fennel, thyme, and lemons over the octopus. Sprinkle over the coriander seeds and salt. Cover the pan, place it in the oven, and cook the octopus until fork-tender, about 5 hours. The suckers and outer skin should easily slide off; if not, cook it longer. Allow the octopus to cool in the oil, uncovered, at room temperature for 1 hour.

Remove the octopus from the oil, draining it well; reserve the oil. Remove the suckers and outer skin. Cut off the legs. Rinse the head and cut it into pieces the same diameter as the legs. Strain the oil, pour it into a container, and add the pieces of octopus. Cover and refrigerate. *(The octopus can be refrigerated for up to 4 days.)*

FOR THE EGG YOLKS: Preheat the water bath in an immersion circulator to 71°C (see sidebar, page 118).

Place the egg yolks in a vacuum bag and seal it at the highest setting. Cook the yolks in the water bath at 71°C until cooked through, about 18 minutes.

While the yolks are cooking, make an ice bath in a bowl with equal parts ice and water. When the yolks are cooked, put the bag in the ice bath until they are cool, about 10 minutes. Remove the yolks from the bag and pass them through a fine sieve into a bowl. Season with salt. Transfer to a piping bag fitted with a ¼-inch plain tip until ready to use. *(The yolks can be refrigerated for up to 1 day.)*

WHEN READY TO GRILL THE OCTOPUS: Prepare a very hot grill. Allow the grill grates to preheat so that the octopus won't stick.

MEANWHILE, FOR THE PARSNIPS: Put the parsnip puree and squid ink in a food processor and emulsify, about 2 minutes. Season with salt and white pepper. The puree can be kept warm in a covered small saucepan over low heat for up to 20 minutes.

TO GRILL THE OCTOPUS: Season the octopus with salt. Place it over the hottest part of the grill, with the grill lid open, and char on both sides, about 3 minutes per side.

MEANWHILE, FOR THE FENNEL: Chop enough fronds to get 1 tablespoon. Cut the bulb into 2-inch-long batons. Combine the fennel, fronds, olive oil, fennel extract, and salt in a bowl. Toss with the coriander, add the lemon juice, and toss again.

TO COMPLETE: Divide the octopus among six plates. Place dollops of the parsnip puree on each. Place some fennel next to the puree. Pipe out a dollop of egg yolk on each plate. Make an indentation in the yolk and fill it with olive oil. Sprinkle with fennel pollen, fronds, and flowers, and with coriander.

FLOUNDER CRUDO
WITH RHUBARB, BUTTERMILK, RADISHES, AND SEA BEANS

SERVES 4

We are lucky enough to get flounder straight from the boat in Charleston. Once the rigor mortis goes away and the muscles of the fish have had some time to relax, I love serving the delicate flesh totally raw. The flavor is stunning. For this dish, the rhubarb serves to add a definite sourness and the raw radish offers a nice little spicy bite. The buttermilk is an acidic fat that brings the dish together, while the sea beans add crunch, salt, and that wonderful flavor of seawater. It all adds up to a very refreshing dish.

1 cup whole-milk buttermilk, preferably Cruze Farm Buttermilk (see Resources, page 326)

1 teaspoon Espelette vinegar (see Resources, page 326)

1 teaspoon piment d'Espelette (see Resources, page 326)

One 7-ounce flounder fillet, skinned and any bones removed

Grated zest (use a Microplane) and juice of 1 lime

Sea salt, preferably Charleston or Maldon (see Resources, page 326)

1 tablespoon extra-virgin olive oil

4 medium breakfast radishes, shaved paper-thin

20 radish flowers

2 rhubarb stalks, peeled and cut into small dice

4 ounces sea beans, blanched in boiling water to remove some of the saltiness, drained, and cut into ½-inch-wide slices

Chill four plates. Combine the buttermilk, vinegar, and piment d'Espelette powder in a small bowl. Cover and refrigerate.

Use a long, thin knife to slice the flounder as thin as possible without tearing it: Start with the heel of the knife and pull it toward you in one long, fluid motion (no sawing!). Arrange the slices on the chilled plates.

Sprinkle the lime zest over the fish and drizzle the lime juice over. Season the fish with sea salt. Drizzle the buttermilk mixture around the fish. Lightly sprinkle the olive oil over the fish. Garnish with the radish slices, flowers, diced rhubarb, and piment d'Espelette. Sprinkle the sea beans over the fish and buttermilk and serve immediately.

PANFRIED SHAD ROE
WITH CREAMED RICE, COLLARD GREEN SAUERKRAUT, AND BACON JAM

SERVES 6

Shad roe is a coveted item in the South, and the season is very short. Whenever the dog-woods are in bloom, the roe begins showing up in the market, and you are sure to find it in my kitchens. The first time I had shad roe, it was wrapped in bacon and served over lem-ony spinach. I fell in love with it, and every time the shad run begins, I can taste that first expe-rience in my head as plain as yesterday. This dish is a nod to that shad roe. Be careful when cooking the roe sacs—they like to pop and crackle. Use a mesh grease splatter guard to protect yourself, and don't let the hot oil scare you away from eating one of the South's most treasured delicacies.

Note that the collard kraut must ferment for at least 7 days.

COLLARD KRAUT

3 large bunches collard greens

7 tablespoons kosher salt

1½ teaspoons crushed red pepper flakes

Juice of 1 lemon

BACON JAM

13 ounces bacon, preferably Benton's (see Resources, page 326), cut into ¼-inch dice

2 tablespoons brown sugar

⅔ cup sorghum (see Resources, page 326)

½ cup plus 2 tablespoons Chicken Stock (page 318)

¼ cup sherry vinegar

1 tablespoon soy sauce

RICE

1½ cups Anson Mills Carolina Gold Rice Grits (see Resources, page 326)

6 cups milk

1 fresh bay leaf

2½ teaspoons kosher salt

½ cup freshly shredded Pecorino Romano

SHAD ROE

6 small shad roe sacs (about 4 ounces each)

Kosher salt

1 cup Anson Mills Rice Flour (see Resources, page 326)

¼ cup canola oil

4 tablespoons unsalted butter

FOR THE KRAUT: Without removing the ribs, make stacks of collard leaves, roll them into cylinders, and cut them into thin ribbons. Wash the collards in a sink or large bowl of cold water, changing the water several times if the collards have sand on them. Drain and dry.

Mix the salt, red pepper flakes, and lemon juice together in a large bowl or nonreactive container. Add the collards and mix well. Weight the collards down with a plate to keep them submerged in the liquid that will be released. Cover the bowl with cheesecloth and tie it snugly. Let the collards sit in a cool, damp place for at least 7 days to ferment.

Transfer the kraut to a glass or plastic container and refrig-erate until ready to serve. *(Tightly covered, the kraut will keep for at least 2 weeks in the refrigerator.)*

FOR THE BACON JAM: Put the bacon in two skillets large enough to hold it in one layer and cook over medium-low heat, stirring frequently, until the fat is rendered and the bits of bacon are crispy, 4 to 5 minutes. Using a slotted spoon, remove the bacon bits to a paper towel to drain. (Reserve the fat for another use, if desired.)

Combine the brown sugar and sorghum in a medium saucepan and cook over medium heat, stirring only until the sugar dissolves. Then continue to cook, without stirring, gently swirling the pan occasionally to help the mixture cook evenly, until it changes color and caramelizes, about 5 minutes. Add the stock and vinegar, being very careful, as it will splatter. Bring the mixture to a simmer and cook until it has reduced by half, about 10 minutes.

Add the soy sauce and bacon and bring to a simmer; the mixture should be thick and sticky. Set aside at room tempera-ture. *(Tightly covered, the bacon jam will keep for up to 2 days in the refrigerator.)*

FOR THE RICE: Put the rice and milk in a large saucepan, add the bay leaf and salt, and bring to a simmer over medium heat. Reduce the heat to low and cook the rice, stirring frequently to prevent sticking, until it is tender, about 20 minutes. Remove the bay leaf and stir in the cheese and a splash of lemon juice. Cover and keep warm on the back of the stove for up to 20 minutes.

FOR THE SHAD ROE: You need to cook the roe in two skillets, or in two steps; if you only have one cast-iron skillet, you can hold the first round in a 200°F oven for 10 minutes. Heat two large cast-iron skillets over high heat. Meanwhile, season the roe sacs with salt. Put the rice flour in a medium bowl and gently coat the roe with it.

When the skillets are hot, add the canola oil. When the oil shimmers, gently lay the roe in the hot oil (be careful, the oil

will sizzle and splatter) and reduce the heat to medium. Use a splatter guard if you have one; if a roe sac gets a hole in it, it will pop and explode. Cook the roe until golden brown on the first side, about 3 minutes. Gently turn it over, add the butter to the skillets, and cook the shad roe until red in the center, about 3 minutes.

TO COMPLETE: Divide the rice and roe among six large warm plates. Top each with a small pile of kraut and a dollop of bacon jam.

NOTE

The recipe makes more collard kraut than you will need for this dish, but it keeps well and you can use it anywhere you'd use regular sauerkraut, such as with hot dogs or sausages.

CORNMEAL-DUSTED SNAPPER

WITH BREAD-AND-BUTTER COURGETTES AND RED PEPPER BUTTER SAUCE

SERVES 4

The snapper/grouper boats operating out of Charleston bring in a bountiful harvest every season, and the town overflows with prime fresh seafood. Mark Marhefka provides us with most of our snapper, and this is one of my favorite ways to serve it. Every country boy knows how to dust a snapper fillet with cornmeal and drop it in hot grease, but we fancy it up with a butter sauce that brings with it the sunshine-flecked goodness of a ripe red summertime bell pepper and with the sharp tang of pickled squash. It's a classic fish camp feast, dressed up just a little.

Note that the pickling liquid for the squash has to steep overnight in the refrigerator.

COURGETTES

1 cup Vegetable Stock (page 316)

2 cups ⅛-inch dice white onion

½ jalapeño pepper, seeded and shaved paper-thin

¼ cup cider vinegar

¼ cup sugar

1 teaspoon kosher salt

1 teaspoon yellow mustard seeds

½ teaspoon celery seeds

½ teaspoon turmeric

2 whole cloves

1 fresh bay leaf, torn into pieces

2 medium yellow squash, shaved into thin disks on a mandoline

2 medium zucchini, shaved into thin disks on a mandoline

4 squash blossoms, cut into very thin strips

RED PEPPER BUTTER SAUCE

5 red bell peppers (about 3 pounds), cored, seeded, and cut into strips

1 teaspoon rice wine vinegar

2 tablespoons unsalted butter, diced and chilled

SNAPPER

Four 7-ounce skinless snapper fillets, ¾ to 1 inch thick

Kosher salt

Cayenne pepper

1 cup cornmeal, preferably Anson Mills Antebellum Fine White Cornmeal (see Resources, page 326)

Canola oil

Squash blossom leaves and garlic flowers

EQUIPMENT

Juice extractor

FOR THE COURGETTES: Pour the vegetable stock into a large saucepan, add the onion, jalapeño pepper, vinegar, sugar, salt, mustard seeds, celery seeds, turmeric, cloves, and bay leaf, and bring to a simmer over high heat. Remove from the heat, transfer this pickling liquid to a container, and cool to room temperature. Cover and refrigerate overnight to steep.

THE NEXT DAY, MAKE THE RED PEPPER BUTTER SAUCE: Run the red pepper strips through a juice extractor. Strain the juice through a fine-mesh sieve.

Put the juice into a medium nonreactive saucepan and bring to a boil over high heat, skimming off any scum that rises to the top. Reduce the heat to medium and cook the juice at a simmer until it is reduced to a syrup, about 30 minutes.

Stir the vinegar into the syrup, being sure to incorporate it well. Reduce the heat to low and add the butter piece by piece, stirring until it is emulsified. Use immediately, or serve at room temperature. (The red pepper butter sauce can be kept at room temperature for up to 2 hours; it should not be reheated.)

FOR THE SNAPPER: Season the fish with salt and cayenne pepper. Put the cornmeal in a shallow bowl and season it with salt. Dredge the fillets in the cornmeal, gently shaking off any excess.

Heat a large skillet over high heat. When the skillet is hot, add ¼ inch of canola oil. When the oil shimmers, add the fillets skin side up. Do not shake the skillet or touch the fillets for 1 minute. Reduce the heat to medium-high and cook until the fillets are golden brown on the first side (peek under them to check), 3 to 5 minutes. Turn them over and repeat the process. Remove them from the skillet and blot on a paper towel. The fillets can be kept warm in a 200°F oven for 5 minutes.

TO FINISH THE SQUASH: Bring the pickling liquid to a simmer in a deep skillet over medium-high heat. Add the yellow squash, zucchini, and squash blossoms, cook for 1 minute, and drain.

TO COMPLETE: Place a snapper fillet in the center of each of four large, warm plates. Ladle ¼ cup of the red pepper butter sauce near the fish on each plate. Place a small pile of the courgettes at either end of each fillet. Garnish with squash blossom leaves and garlic flowers.

GRILLED COBIA
WITH PEANUT ROMESCO AND EGGPLANT BARIGOULE

SERVES 6

The cobia is one mean-looking fish, sort of a cross between a sleek submarine and an indestructible torpedo. Its fat-to-meat ratio is just wonderful, allowing it to stand up to the hot coals of a summertime grill. The outside caramelizes a little bit and the fatty interior is a great reward. This dish is deceptively simple, with just a Spanish romesco sauce and a little eggplant on the side. Don't let it fool you, though; it's packed with flavor.

Note that the peanuts must be soaked overnight.

PEANUT ROMESCO

½ cup extra-virgin olive oil

1 dried ancho chile

1 medium Cherokee Purple Tomato or other heirloom tomato with dark, meaty flesh

1 large red bell pepper

1 small sweet onion, unpeeled

½ cup raw peanuts, soaked overnight in cold water

1 garlic clove

2 tablespoons sherry vinegar

1 slice whole wheat bread, toasted and cut into ½-inch cubes (to make about ½ cup)

1 teaspoon smoked paprika (see Resources, page 326)

Kosher salt

COBIA

Twelve 3-ounce skinless cobia fillets, about 2 inches thick

4 tablespoons extra-virgin olive oil

Kosher salt

Piment d'Espelette (see Resources, page 326)

Eggplant Barigoule (page 47), warmed

12 small dandelion leaves

FOR THE PEANUT ROMESCO: Heat 1 tablespoon of the olive oil in a small skillet over medium-high heat. Add the ancho chile and fry, turning once, until darkened and slightly puffed, about 30 seconds. Transfer the chile to a small bowl, add enough hot water to cover, and let stand for 30 minutes.

Meanwhile, preheat the broiler to high. Put the tomato, bell pepper, and onion in a baking dish and toss with 2 tablespoons of the olive oil. Broil, turning frequently, for about 8 minutes, until partially charred on all sides. Cover with foil and let stand for 15 minutes.

Peel and seed the tomato and bell pepper and place them in a blender. Peel the onion, coarsely chop, and add to the blender.

Heat 1 tablespoon of the olive oil in small skillet over medium-high heat. Add the peanuts and ancho chile and cook until lightly toasted, about 1 minute. Add the peanuts to the blender, then add the remaining ¼ cup olive oil, the garlic, vinegar, bread, and paprika and blend to a coarse puree, about 5 minutes. Transfer the mixture to a bowl and season with salt. Cover and set aside for up to 2 hours. *(Tightly covered, the romesco can be kept for up to 4 days in the refrigerator. Let it come to room temperature before serving.)*

FOR THE COBIA: Prepare a very hot grill. Allow the grill grates to preheat so that the fish won't stick.

Coat the fillets with the olive oil. Season with salt and piment d'Espelette. Cook the fish over the hottest part of the grill, with the grill lid open, for about 5 minutes, without moving the fillets. When they can be easily moved without any resistance, turn them over and cook until about medium-rare, an additional 5 to 7 minutes. (You can cook for 2 more minutes for medium, but cobia will be dry if cooked to well-done.)

TO COMPLETE: Place the cobia fillets off center on each of six warm plates. Place a tablespoon-sized dollop of the romesco at the top and at the bottom of each fillet. Add 2 quenelles (oval-shaped scoops) of barigoule to each plate, one near the fish and one off to the side. Garnish the plates with the dandelion greens.

SWORDFISH
WITH CELERIAC ROASTED IN HAY, CIDER-BRAISED LETTUCE RIBS, AND COUNTRY HAM EMULSION

SERVES 6

It's hard to beat a perfectly grilled piece of fish, especially when you're cooking over a hardwood charcoal fire. And there's nothing much better to grill than a super-fresh and fatty piece of swordfish. Coupling grilled swordfish with vegetables roasted in hay is a perfect match. I roast a number of different vegetables, especially root vegetables, this way. The hay acts as an insulator while adding a unique, earthy flavor. Using hay as a cooking medium isn't new. In earlier times, meats were often packed in hay as a means of keeping them warm before service, and I imagine that it wasn't long before the flavoring mechanism was noticed and moved into the kitchen. A technique like this lets us use parts of a plant that would otherwise go into the trash—here the hay adds a grassy, earthy flavor—which is something I love to do. It's important to look at plants with the same approach we use when butchering whole animals. We try not to throw anything away, hence the lettuce stems in this recipe. Stems from very mature plants have an amazing texture and flavor. We finish the dish with a sauce that really captures the essence of our prized country hams here in the South; it's a great way to get the flavor of the ham in a new way.

COUNTRY HAM EMULSION

½ cup diced smoked country ham, preferably Benton's (see Resources, page 326)

2 cups heavy cream

16 shavings katsuobushi (see Resources, page 326)

1 teaspoon apple cider vinegar

¼ cup rendered country ham fat (see Note)

CELERIAC

1½ ounces Timothy hay (see Note)

2 tablespoons unsalted butter, diced

2 celeriac (about 1 pound each), scrubbed but not peeled

Kosher salt

LETTUCE RIBS

36 radicchio ribs

2 cups apple cider

½ tablespoon agave nectar

1 tablespoon cider vinegar

1 teaspoon kosher salt

SWORDFISH

Six 6-ounce swordfish steaks, about 1½ inches thick

2 tablespoons extra-virgin olive oil

Kosher salt

Piment d'Espelette (see Resources, page 326)

FOR THE COUNTRY HAM EMULSION: Put the ham and cream in a medium saucepan and bring to a simmer over high heat. Lower the heat to medium and cook until the cream has reduced, about 15 minutes. Add the katsuobushi, remove from the heat, and let steep for 2 minutes.

Strain the cream through a fine-mesh sieve into a small saucepan. Add the cider vinegar and ham fat and use an immersion blender to emulsify. Keep warm at the back of the stove.

FOR THE CELERIAC: Preheat the oven to 400°F. Lay out a 2-foot square of aluminum foil on your work surface (use two or three overlapping sheets if necessary). Make a nest with (half) of the hay on the foil and scatter the butter over the hay. Season the celeriac with salt, place it on top of the hay, and pile the rest of the hay over it. Bring the edges of the foil together to make a pouch with an opening at the top. Place the pouch on a rimmed baking sheet and bake the celeriac for about 30 minutes, or until fork-tender.

Meanwhile, prepare a very hot grill. Allow the grill grates to preheat so that the fish won't stick.

WHILE THE GRILL IS HEATING, PREPARE THE LETTUCE: Combine the apple cider, agave nectar, cider vinegar, and salt in a medium saucepan and bring to a simmer over high heat. Add the radicchio ribs, cover, reduce the heat to low, and cook the ribs until tender, about 10 minutes. Keep warm at the back of the stove.

FOR THE SWORDFISH: Coat the steaks with the olive oil. Season with salt and piment d'Espelette. Cook the swordfish steaks over the hottest part of the grill, with the grill lid open, for about 5 minutes, without moving them. When the steaks can be easily moved without any resistance, turn them over and cook for an additional 5 minutes, or until the flesh just begins to flake.

TO COMPLETE: Remove the celeriac from the foil and hay and cut each into 6 wedges. Use an immersion blender to froth the ham emulsion by tilting the saucepan to one side and allowing the bubbles to collect on the other side.

Place a swordfish steak in the center of each of six warm bowls and place 2 wedges of celeriac next to the swordfish. Place 6 radicchio ribs around the fish and top with about ¼ cup of the froth from the country ham emulsion.

================ NOTE ================

For rendered country ham fat, we cut away the fat from the outside of a ham and cook it in a skillet over low heat until all the fat is rendered and the ham is crispy. *One cup diced country ham will produce the ¼ cup fat you need for this recipe.*

You can buy Timothy hay online (see Resources, page 326).

SEED-CRUSTED SNAPPER WITH ROASTED OKRA, TOMATO DASHI, AND BENNE

SERVES 6

I often use a vegetable-and-herb-seed crust to add a little personality and texture to seared fish. This one can be used in a lot of different dishes, as the flavors work well with a multitude of foods. Okra sure does get a bad rap because of its sometimes mucilaginous texture. But if you don't let okra come in contact with water, the slime-producing reaction won't occur. So roasting okra, as here, is a great way to convert any okra haters out there.

I'm obsessed with katsuobushi, a Japanese dried, fermented, and smoked tuna product. A trip to Japan always influences my cooking. Although the cuisine is thin on meat products, at least relative to the United States, the deep umami flavors that the Japanese create resonate with me deeply. I swear, sometimes high-quality katsuobushi tastes like Benton's country ham or smoked bacon. I like to have people close their eyes, smell it, and try to guess what it is. Everyone always says bacon or ham, so it fits into our cuisine quite well. If you think of the tomato dashi as something akin to a country ham and tomato broth, then it makes perfect sense in a new South.

DASHI

5 very ripe heirloom tomatoes (about 3 pounds), cored and cut into chunks

3 sheets silver-strength gelatin (see Resources, page 326)

12 shiitake mushrooms with stems, thinly sliced

One 2-by-4-inch sheet of kombu (see Resources, page 326), wiped clean

½ cup katsuobushi shavings or dried bonito flakes (see Resources, page 326)

SEED CRUST

⅔ cup fennel seeds

⅓ cup anise seeds

¼ cup onion seeds

3½ tablespoons basil seeds

3 tablespoons coriander seeds

TOMATOES

30 Sungold or other cherry tomatoes

OKRA

2 tablespoons canola oil

12 large okra pods, cut lengthwise in half

Juice of 1 lemon

Pinch of togarashi chile flakes (see Resources, page 326)

1 tablespoon extra-virgin olive oil

SNAPPER

Canola oil

Twelve 3-ounce skinless snapper fillets, 1 to 1½ inches thick

Kosher salt

GARNISH

½ cup Anson Mills Antebellum Benne Seeds (see Resources, page 326)

Okra shoots or flowers (optional)

EQUIPMENT

Butcher's twine

FOR THE TOMATO DASHI: Puree the tomatoes in a food processor. Remove ½ cup of the puree and leave the remaining puree in the processor.

Place the gelatin sheets and ½ cup cold water in a shallow bowl and let the gelatin soften for 2 minutes. Meanwhile, warm the ½ cup pureed tomatoes in a small saucepan over medium-low heat.

When the gelatin sheets are soft, lift them from the water and gently wring them out. Add them to the warm tomatoes and heat, stirring, until dissolved, about 1 minute. Add the dissolved gelatin mixture to the remaining pureed tomatoes in the food processor and pulse a few times to combine. Pour the tomato puree into ice cube trays and freeze until frozen solid.

Set a strainer over a deep bowl and line it with a double layer of cheesecloth. Put the frozen tomato cubes in the strainer, gather up the edges of the cheesecloth, and tie with butcher's twine to form a sack. Attach the sack to a dowel or the handle of a wooden spoon, remove the strainer, and suspend the sack over the bowl. Refrigerate overnight. Clear liquid (tomato water) will drip into the bowl for about 24 hours; if any red liquid begins to appear, remove the sack (do not squeeze the cheesecloth, or the tomato water will cloud).

Pour the tomato water into a large nonreactive saucepan, add the mushrooms, and bring to a simmer over high heat. Cover the pan, reduce the heat to medium, and simmer the mushrooms for 20 minutes to extract their flavor.

Make an ice bath in a bowl with equal parts ice and water. Pour the mushrooms and broth into a bowl, set the bowl in the ice bath, and stir until the broth cools to 50°F.

Strain the broth into a medium nonreactive saucepan, add the kombu, and heat over medium heat until the broth reaches a temperature of 140°F. Keep the mixture steady at 140°F for 1 hour by using a digital thermometer and adjusting the heat as necessary.

Remove the kombu, increase the heat under the broth to medium-high, and bring the broth to 176°F. Remove the pan from the stove, add the katsuobushi shavings, and allow them to rehydrate, about 15 seconds. Leave them for 90 seconds more, then strain the liquid through a fine-mesh sieve into a medium nonreactive saucepan, without pressing down on the katsuobushi; discard the shavings.

The dashi can stand at room temperature for up to 30 minutes. Or transfer it to a container, cool it to room temperature, cover, and refrigerate it until ready to use. *(Tightly covered, the dashi will keep for up to 3 days in the refrigerator; it can also be frozen for up to 1 month.)*

FOR THE SEED CRUST: Grind all of the seeds in a mortar and pestle. You just want to crack them into smaller pieces, not

make a powder. *(The cracked seeds will keep in a tightly covered container, in a cool, dark place, for up to 2 weeks.)*

FOR THE TOMATOES: Bring a small saucepan of water to a boil. Cut a small slit in the top of each tomato. Place a colander or strainer in the sink and start running cold water into it. Add the tomatoes to the boiling water and blanch for 5 seconds, then pour the tomatoes and water into the strainer and allow the cold water to cool them. Drain and peel off the skin, using the tip of a paring knife if necessary. Put the tomatoes on a plate and set aside.

FOR THE OKRA: Heat a large cast-iron skillet over high heat. Add 1 tablespoon of the canola oil. When the oil shimmers, add half of the okra cut side down, and cook until seared and golden brown on the cut side, about 2 minutes. Turn the okra over and cook for 2 minutes more. Transfer the okra to a plate covered with a paper towel. Wipe out the skillet, add the remaining tablespoon of canola oil, and repeat the process to cook the second half of the okra.

Return all the okra to the skillet, add the lemon juice, togarashi, and olive oil, and toss to coat.

The okra can be held, covered, on the back of the stove for up to 10 minutes.

FOR THE SNAPPER: Heat two large cast-iron skillets over high heat, and add enough oil to cover the bottom of the skillets. Season the fillets with salt and coat the skin side with the seeds, using about half of the seed mixture (reserve the rest for another dish). When the oil shimmers, add 6 pieces of fish to each skillet seed side down, and sear them for 2 minutes without shaking the skillets or touching the fillets. Reduce the heat to medium-high and cook until the fillets are golden brown on the first side (peek under the fillets to check), about 4 minutes. Turn them over, remove the skillets from the stove, and allow the fillets to cook off the heat for another 30 seconds, then remove from the skillets.

Meanwhile, reheat the dashi over medium heat until warm.

TO COMPLETE: Place 2 snapper fillets in each of six warm rimmed soup plates. Place 5 Sungold tomatoes around the fish in each bowl and stand 4 okra halves around the fish. Sprinkle the tomatoes and okra with the benne seeds. Pour a generous ¼ cup dashi into each bowl. Garnish with okra shoots or flowers, if you have them.

MARK MARHEFKA

Mark Marhefka can discuss the intricacy of federal fisheries regulations as fluently as a Washington, D.C., bureaucrat and then transform into a fearless man of the sea as soon as his boat leaves the dock. He understands the art and the practice of commercial fishing like no other man on earth.

I had heard about Mark from a couple of chefs when he decided to start selling his fish straight from the dock. This was exciting news for all of us in Charleston; it was a real game changer. We knew the quality of the fish we'd be cooking would improve dramatically.

Now, a decade later, we are all spoiled rotten. I used to think that I knew what fresh fish was. I sometimes ordered fish direct from Hawaii, airmailed packed in ice. I was proud of the product I was using. And then came Mark. It's hard to describe what these moments in your career feel like. One part of you is overjoyed and the other half of you is embarrassed and angry. You're happy because you've discovered a new way forward, one that promises tremendous results, and you're angry because it feels like people have been dishonest with you for years. "This one is straight off the boat, chef, as good as it gets!" Yeah, right. But I love those moments when you realize that there is a better way. Straight from the boat is one of them.

Of course the fish is fresher, but it's also sustainable. Mark doesn't mine the best-tasting catch in the ocean at the lowest price of extraction. He doesn't haul in a thousand pounds of flesh on a long line

of hooks that also wastes three thousand pounds of bycatch. He operates on single lines, bottom-fishing for the snapper/grouper category, as it's called, but he also functions as an educational director, teaching us what's edible at the bottom of the sea. He introduced me, and a host of other local chefs, to triggerfish, rudderfish, white grunt, amberjack, and scores of other underused varieties. When one variety is overfished, or put under regulatory protection, some fishermen bend the rules or lie. Mark instead changes course and brings in something you've never heard of before, and it's always delicious. He has single-handedly transformed the culinary culture of Charleston.

He has been able to do this because he runs a sustainable operation. He offers what he calls a Community-Supported Fishery to retail customers, which is similar to the widely popular CSAs (Community Supported Agriculture) that dot the modern culinary landscape. His CSF is so popular that the waiting list is now impossibly long.

You may not have fresh fish where you live, but here is the best advice I can give. It doesn't matter if it's a fish or a carrot, get to know the person who supplies the food you eat. Become friends with him or her. Mark and I are great friends. We eat dinner together and pass around jars of moonshine, and we even hang out when our favorite bands come to town. I trust Mark. That's important when it comes to food. Would you give your friend old fish?

THE LARDER

SAVING UP

Pickling and preserving are two techniques that anchor Southern cooking. Putting up food can be one of the most rewarding things you do in your kitchen. There is something special and archetypal about the annual rituals of cleaning vegetables, making a solution to preserve them, and processing the jars through crucibles of heat and pressure. Even the routine of cleaning the jars brings me a greater sense of meaning and purpose—it makes me feel like I'm doing something 100 percent honest.

My memories of growing up include Mason jars, crocks of fermenting mixed vegetables, seeds drying on windowsills in autumn, and hand-rubbed country hams hanging to cure through the winter. These traditions shaped me into the person I am today, and continuing them is central to my purpose as a chef. But they are also essential to authentic Southern cooking. The larder is the backbone of a true Southern cook's kitchen. In a part of the country where most families have come from poverty at some point, and everyone knows the draw of a well-set table for Sunday supper, the pantry represents the sum total of a cook's efforts throughout the year. My cooking has been dedicated to building a larder, and it's been an exciting journey, full of discoveries and deliciousness.

My family always made pickles. Just about every meal we ate had some sort of pickled element, and now acidity is a big part of my cuisine. The nice thing about pickling is that it is very easy to do. A pickle can be as simple as adding vinegar to a vegetable and letting it sit for a few days. If you want to get fancy, add some salt and sugar to taste. If the vegetables are cut thin enough, they'll be pickled in no time.

I like to go to the grocery store for inspiration for stocking my pantry—the more generic the store, the better. I peruse the shelves, thinking about the origins of food and taking in all the varieties of jams, vinegars, oils, condiments, and spices. That is how I get inspired to make my own. Real food can be both cheap and convenient. Want to make real vinegar for cheaper than what you'll pay in the grocery store? It's easy. Mix one part "live" vinegar (the kind you find at the fancy grocery store; it sticks out because it has a bunch of weird stuff floating around in it) in a jar with three parts leftover booze—wine, "skunked" beer, whatever you have. Seal the jar, wrap it in a towel, set it in a cool, dark place, and wait about a month or so. Just like magic, you'll have made your own vinegar from your favorite wine, beer, or spirit. That doesn't sound difficult, does it?

One of my favorite pickles involves an old technique that relies on a group of bacteria called lactobacillus. It's the same stuff that gives yogurt and buttermilk their soulful tang. My grandmother used this technique to make her "mixed pickles" of corn, green beans, and cabbage. I still have a couple of jars in my garage that she made, but I don't reckon I'll ever eat them. I keep them as a reminder of the things she taught me when she was alive. The jars still have her gorgeous handwriting on the lids.

My grandmother's technique for mixed pickles is very easy. Take a big old crock or glass container, fill it a little over halfway with water, and then carefully place an egg (uncooked, still in the shell) in the water. Add kosher salt until the egg floats and starts to break the surface. This makes the pickle pretty darn salty, which I don't mind, but people who are sensitive to salt shouldn't add quite as much. Clean your vegetables: snap the beans in half, grate the cabbage, remove the corn kernels from the cobs. Then blanch the vegetables in a pot of boiling water and shock (cool) them in an ice bath. Put the vegetables in an old pillowcase—make sure you rinse it a couple of times to be sure there isn't any soap in it—and put the bag into the brine. It's important that the vegetables stay submerged. Most people today fill Ziploc bags full of more brine (in case they come open) and put them on top to weight down the vegetables. Old-timers (like me) put a creek rock on top of the pillowcase as a weight. Put the lid on top of the crock and let it sit in a cool basement or garage for a month. The lactobacillus that is naturally present on the vegetables will eventually produce lactic acid and preserve the vegetables. If you've ever had sauerkraut, you've eaten food made with the same technique.

Good cooking is all about building up your pantry and viewing it as your arsenal. A well-stocked pantry or a fridge full of pickles can add diversity to your cooking. Plus, you'll never be caught empty-handed for a last-minute dinner again.

SOME EASY RULES
FOR PICKLING, PRESERVING,
AND PUTTING UP

IN THIS CHAPTER, YOU WILL FIND THREE LEVELS OF "putting up." The first is quick pickling, making pickles and relishes that are kept in the refrigerator. The second is preserving, jams, jellies, preserves, and marmalades that are processed so that they can be safely kept in a pantry or other cool, dark place for a finite period of time. The third is canning, which is for the long term.

If you want to put up fruits and vegetables to preserve the bounty of a season, as my grandmother filled her larder, there are specific safety rules you must follow. For this type of canning, the best understanding comes from a good study of the detailed instructions provided by sources such as the National Center for Home Food Preservation (http://nchfp .uga.edu/how/can_home.html) or the Ball company (www .freshpreserving.com/getting-started.aspx). I encourage you to seek out these sources and others.

Here are some tips to get you started, whether you're making a quick pickle or canning for future use.

- White vinegar and apple cider vinegar are your acids of choice. You can experiment with vinegars for quick pickles because they will be refrigerated, but for preserving or

canning safety, you must use a vinegar with an acidity of 5 percent. The acidity acts both to preserve and to protect the food from bacteria.
- Use kosher salt or pickling salt. Table salt with iodine or sea salts with minerals will alter the flavor.
- For quick pickles, your container or jars should be clean and dry, but they do not have to be sterilized.

For preserving and canning, the jars must be sterilized. Here's how:

1. Fill a large canning pot with a rack three-quarters full of water. Place your canning jars and rings in the pot and place it on the stove over high heat. When the water comes to a boil, set a timer for 5 minutes.
2. Meanwhile, heat the lids in a saucepan of hot water to 180°F. Never boil the lids, because the sealant material may get damaged and won't produce a safe seal.
3. Once the jars and rings have boiled for 5 minutes and the water in the pan of lids has reached 180°F, remove the pot and pan from the stove and cover them. Have a clean kitchen towel ready. When you are ready to fill the jars,

using canning tongs, remove them from the pot and invert them onto the kitchen towel. Leave them there for 1 minute before you turn them over and fill them. You want the jars to be hot when you put the food into them. Use tongs to remove the lids and rings and shake off any water before putting them on the jars.

PRESERVING

- Ladle the hot preserves into the hot sterilized jars, leaving a ¼- to ½-inch headspace (the space between the top of the food and the top of the jar), as directed.
- Wipe the rims and threads of the jars clean with a damp paper towel.
- Attach the sterilized lids and rings and tighten the rings.
- Cool on a dish towel or a rack, not directly on the countertop.

FOR BOTH PRESERVING AND CANNING

It is important that the jars seal properly and a vacuum forms. When they do, the lids become concave in the center and you can't pop the button up and down. The lids usually ping when this happens, but since sealing may take several hours, you should always check the lids before you store the jars, not rely on hearing a ping. If the center is firm and will not pop up and down when pressed, it is safe to store the item in the pantry or other cool, dark place for up to 6 months. If the jar has not sealed properly, store in the refrigerator and use within 2 weeks.

─── TIP ───────────────────────
Widemouthed jars are the easiest to fill. If you didn't inherit your grandmother's widemouthed canning funnel, buy one—it makes ladling the preserves into the jars so much easier. You can reuse the rings as well as the jars, but do not reuse lids. They will only seal properly once.

BOILING-WATER-BATH CANNING

A boiling-water canning pot must have a fitted lid and a removable rack to hold the jars, preferably one with handles so that all the jars can be loaded into the rack and lifted into and out of the canning pot at once. It must have a flat bottom if you have an electric or induction stovetop, and it should be no larger than 4 inches wider in diameter than the burner or heat element, no matter what type of stove you have. The pot must be deep enough that the jars will always be covered with at least 1 inch of boiling water during processing.

- Ladle the food into the sterilized jars through a widemouthed funnel, making sure to leave the headspace specified in the recipe.
- Run a thin-bladed knife around the inside of the jars to release any air bubbles. Wipe the rims and threads clean. Attach the lids and rings. Screw the rings on firmly but not tight ("finger-tighten").
- Have the water in the canning pot at a high boil. Lower the jars into the boiling water. If using canning tongs, hold the jars below the rings and be careful not to tilt them. The jars must be covered by at least 1 inch of boiling water at all times during processing; add more boiling water if needed. Cover the pot and return the water to a boil. Process the food for the time specified in the recipe.
- Lift the jars out with the rack or canning tongs. Set them at least an inch apart on a clean dish towel or a rack, not directly on the countertop. Cool to room temperature and make sure the jars have sealed before tightening the rings again and storing.

PICKLED EGGS

MAKES 12
PICKLED EGGS

When I was very young, I was terrified of pickled eggs. Looking back, I'm sure it was the unnatural pink dye of commercial varieties, and the way they floated around in the jar like some crazy science experiment. These days, I could easily put down a dozen, although I wouldn't recommend it. We keep these pickled eggs stocked behind the bar at Husk. If you've been to enough country dive bars and rural gas stations and groceries around the South, this should come as no surprise. Once I had a guest tell me that a sign of a good bar was if there were pickled eggs sitting behind the bar. So I've adopted that theory, and I don't plan to disappoint.

12 jumbo eggs

2 pounds beets

3 cups cider vinegar

1 cup white vinegar

½ cup sugar

1 tablespoon plus 1 teaspoon kosher salt

3 whole cloves

½ cinnamon stick

EQUIPMENT

Juice extractor

Using a sewing needle or pushpin, pierce a hole in the shell at the wide end of each egg. Put the eggs in a large saucepan and cover them with room-temperature water. Bring the water to a boil over medium-high heat and boil the eggs for 2 minutes. Remove the saucepan from the stove, cover it, and let the eggs remain in the water for 10 minutes.

Carefully drain the eggs in a colander in the sink, then peel the eggs under cold running water. Put the eggs in a container, cover, and refrigerate.

Run the beets through a juice extractor; you need 2 cups of juice. (If there is extra juice, you may freeze it for a soup or another use.)

Combine the beet juice, cider vinegar, white vinegar, sugar, salt, cloves, and cinnamon stick in a large stainless steel saucepan and bring to a boil over medium-high heat, stirring to dissolve the sugar and salt.

Transfer the mixture to a glass or stainless steel container, let cool, and then refrigerate until completely cool.

Add the eggs, cover, and refrigerate for at least 1 week before eating; stir them occasionally. Tightly covered, the eggs will keep for up to 1 month in the refrigerator.

PICKLED PEACHES

Believe it or not, South Carolina produces an awful lot of peaches, many more than Georgia (even if they do call it the Peach State), so peaches play a big role in the food of the Lowcountry. The season is pretty short. Be careful when buying peaches to pickle: It's essential that you use peaches at their peak of ripeness. If they are not ripe enough, the flavor won't come through, and there's no use in preserving something that's not delicious. But on the other hand, if they are too ripe, they will turn to mush. So cull your basket, eat the ones that are too ripe, and wait for the underripe ones to mature.

These pickles pair well with roasted pork and chicken. I like to serve them with fried green tomatoes and goat cheese as a great little appetizer.

12 ripe peaches

3 cups water

3½ cups white vinegar

2½ cups sugar

1 stalk lemongrass, bruised and chopped

1 tablespoon grated fresh ginger

15 black peppercorns

10 allspice berries

2 whole cloves

1 cinnamon stick

Pinch of ground mace

Bring a large pot filled with water to a boil. Make an ice bath in a bowl with equal parts ice and water. Submerge the peaches in the boiling water for 1 minute. Remove and submerge them in the ice bath to cool them and stop the cooking. Peel the peaches and place them in a glass or stainless steel container.

Bring all of the remaining ingredients to a boil in a stainless steel saucepan over medium-high heat. Reduce the heat to low and simmer for 5 minutes.

Pour the hot pickling liquid over the peaches. Cool to room temperature, then cover and refrigerate for at least 1 week before eating to allow the peaches to cure. Unopened, the peaches will keep for up to 6 months in the refrigerator. Once opened, they will keep for up to 3 weeks longer in the refrigerator.

PICKLED MUSHROOMS

This is a great all-purpose pickle that you can make with every mushroom under the sun. I love having these mushrooms in the kitchen. They add a great acidic punch to many dishes, especially soups and broths.

3 pounds brown beech, cremini, or button mushrooms
1½ cups rice wine vinegar
½ cup apple cider vinegar
1½ cups sugar
½ cup local honey
3 tablespoons Dijon mustard
3 tablespoons whole-grain mustard
1 tablespoon kosher salt
5 thyme sprigs
1 fresh bay leaf

If using brown beech mushrooms that are in clusters, cut off and discard the base of the clusters. Lightly rinse the mushrooms; do not soak them. Dry the mushrooms, cut into bite-sized pieces, and put in a glass or stainless steel container.

Combine all of the remaining ingredients in a small stainless steel saucepan, stir well, and bring to a boil over medium-high heat.

Pour the brine over the mushrooms and cool to room temperature. Cover and refrigerate for at least 1 week before eating to allow the mushrooms to cure. Tightly covered, the mushrooms will keep for up to 2 weeks in the refrigerator.

SPICY PEPPER JELLY

Spicy pepper jelly is pretty common in the South, appearing on the table alongside just about anything fried. I always find the old-fashioned pepper jelly just a little too sweet, so I've toned down the sugar and cranked up the red pepper flavor. If fried foods aren't your thing, try it on a piece of grilled fish.

2 pounds red bell peppers, cored and seeded
2 cups cider vinegar
2 Charleston Hots (peppers; see Resources, page 326) thinly sliced
6 cups sugar
½ cup low-sugar pectin (see Resources, page 326)

EQUIPMENT
Juice extractor
Candy thermometer

Sterilize four pint canning jars, along with the rings and lids (see page 208).

Run the peppers through a juicer; you need 1½ cups juice.

Bring the red pepper juice to a simmer in a medium stainless steel saucepan over medium heat. Skim off any scum. Add the vinegar and hot peppers and bring to a simmer.

Combine the sugar and pectin and whisk them into the pepper mixture and bring the mixture to a boil. Attach a candy thermometer to the pot. Reduce to a simmer and cook until the jelly reaches 245°F, about 10 minutes.

Ladle the hot jelly into the sterilized jars, leaving a ¼-inch headspace. Wipe the rims and threads clean. Place the lids and rings on the jars and tighten the rings. Cool the jars on a clean dish towel or a rack, not directly on the countertop.

It is important that the jars seal properly and a vacuum forms (see page 208). If any jars did not seal, you must store the jelly in the refrigerator. You can eat it immediately, or store it for up to 2 weeks. Properly sealed jars will keep in a cool, dark place for up to 6 months; refrigerate after opening.

SATSUMA ORANGE AND BURNT HONEY MARMALADE

MAKES 4 PINTS

For this recipe, take honey and slowly cook it until it's almost completely burned. Why would you burn honey? I find that the more you cook honey, the less sweet it becomes. And if you take it almost to the point of burning, it acquires a really pleasant bitterness that helps balance the natural sweetness of the satsuma, which gives this jelly a more grown-up, sophisticated flair.

3 pounds Southern satsuma oranges (see Note), washed and dried

1½ cups local honey

2 cups sugar

1 tablespoon low-sugar pectin (see Resources, page 326)

EQUIPMENT

Candy thermometer

Sterilize four pint canning jars, along with the rings and lids (see page 208).

Remove the zest from 1½ of the satsumas with a sharp vegetable peeler. Using a paring knife, carefully remove any white pith from the zest. Cut the zest into very fine strips. Then, with a sharp knife, cut away the peel and white pith from all of the satsumas. Working over a bowl to catch the juices, cut between the membranes to release the segments; reserve. Squeeze the juice from the membranes into the bowl. Discard the membranes.

Put the honey in a small stainless steel saucepan and cook over high heat, stirring vigorously, until it is very dark and has a nice nutty aroma, about 10 minutes. Remove from the heat.

Combine the satsuma zest, juice, segments, honey, and sugar in a medium stainless steel pot and bring to a boil over high heat, stirring to dissolve the sugar. Add the pectin and stir thoroughly to combine. Attach a candy thermometer to the pot, bring the mixture to a boil, and cook, stirring frequently, over medium-high heat until it reaches 245°F, about 15 minutes.

Ladle the hot marmalade into the sterilized jars, leaving a ¼-inch headspace. Wipe the rims and threads clean. Place the lids and rings on the jars and tighten the rings. Cool the jars on a clean dish towel or a rack, not directly on the countertop.

It is important that the jars seal properly and a vacuum forms (see page 208). If any jars did not seal, you must store the marmalade in the refrigerator. You can eat it immediately, or store it for up to 2 weeks. Properly sealed jars will keep in a cool, dark place for up to 6 months; refrigerate after opening.

NOTE

Satsumas other than the Southern satsuma may be used, but they will have a different flavor.

STRAWBERRY–MEYER LEMON JAM

MAKES 3 PINTS

We are fortunate to have an amazing growing season in Charleston. The soil produces some pretty special strawberries, and the first crop comes early in the spring and lasts until the heat of summer sets in. The first pickings are often early enough to catch the last of the Meyer lemons. This combination is intoxicating. When you find a farmer who grows a great strawberry, buy as many as you can and make this recipe so you can enjoy them throughout the year.

3 Meyer lemons, washed and dried

½ cup water

⅛ teaspoon baking soda

7 cups strawberries, hulled and cut into small dice

5 cups sugar

6½ tablespoons apple pectin (see Resources, page 326)

EQUIPMENT

Candy thermometer

Sterilize three pint canning jars, along with the rings and lids (see page 208).

Remove the zest from the lemons with a sharp vegetable peeler. Using a paring knife, carefully remove any white pith from the zest. Cut the zest into very fine strips. Then, with a sharp knife, cut away all the white pith from the lemons. Working over a bowl to catch the juices, cut between the membranes to release the segments; reserve. Squeeze the juice from the membranes into the bowl. Discard the membranes.

Combine the lemon zest, water, and baking soda in a medium stainless steel pot and bring to a boil over high heat. Reduce the heat to low and simmer the mixture until thickened, about 5 minutes; stir frequently to prevent scorching. Add the lemon segments, lemon juice, and strawberries to the pot, increase the heat to medium, and bring back to a simmer. Cover the pot and simmer, stirring occasionally, until the strawberries begin to break down, about 10 minutes.

Add the sugar and pectin to the pot and stir thoroughly to combine. Attach a candy thermometer to the pot, bring the mixture to a boil, and cook over medium heat until it reaches 245°F, about 20 minutes. Stir frequently to prevent the mixture from scorching.

Ladle the hot jam into the sterilized jars, leaving a ¼-inch headspace. Wipe the rims and threads clean. Place the lids and rings on the jars and tighten the rings. Cool the jars on a clean dish towel or a rack, not directly on the countertop.

It is important that the jars seal properly and a vacuum forms (see page 208). If any jars did not seal, you must store the jam in the refrigerator. You can eat it immediately, or store it for up to 2 weeks. Properly sealed jars will keep in a cool, dark place for up to 6 months; refrigerate after opening.

JERUSALEM ARTICHOKE PICKLES

MAKES 4 QUARTS

Linton Hopkins, of Restaurant Eugene in Atlanta, gave me my first jar of homemade Jerusalem artichoke pickles years ago. I went crazy for them and started researching every recipe I could find for these artichoke pickles. What we call Jerusalem artichokes are really the underground corms of a certain sunflower plant. They're also called sunchokes. Linton's gift inspired me to pickle hundreds of pounds of sunchokes that year. It's since become an annual ritual. This recipe is my attempt at re-creating that first jar of pickles that Linton gave to me. The recipe yields 4 quarts, which makes it an ideal recipe to share with friends and family.

1 gallon water

2 cups kosher salt

3¼ pounds Jerusalem artichokes, skin left on, washed, and cut into ½-inch dice

4 cups small dice sweet onion

2 cups small dice red bell peppers

1¾ cups dry mustard

¾ cup all-purpose flour

2 quarts cider vinegar

5 cups sugar

3 tablespoons celery seeds

2 tablespoons freshly ground black pepper

1 tablespoon turmeric

Put the water and salt in a medium stainless steel pot and bring to a simmer over high heat, stirring to dissolve the salt. Cool to room temperature, then refrigerate until thoroughly chilled.

When the brine is chilled, add the artichokes, onions, and peppers, cover, and refrigerate for 24 hours.

Sterilize four quart canning jars, along with the rings and lids (see page 208).

Combine the dry mustard, flour, and 1 cup of the vinegar in a bowl, stirring to make a paste. Combine the remaining 7 cups vinegar, the sugar, celery seeds, black pepper, and turmeric in a large stainless steel saucepan and bring to a boil over medium-high heat. Slowly stir in the mustard paste and simmer the mixture, stirring occasionally, until thickened, about 10 minutes. Remove from the heat.

Drain the vegetables and pack them into the canning jars. Ladle the hot vinegar mixture over the vegetables, leaving a ½-inch headspace. Wipe the rims and threads clean. Place the lids and rings on the jars and finger-tighten the rings.

Process the jars in a boiling-water bath for 20 minutes according to the canning instructions on page 208.

It is important that the jars seal properly and a vacuum forms (see page 208). If any jars did not seal, you must store the pickles in the refrigerator. You can eat them immediately, or store them for up to 3 weeks; the pickles will have more flavor if you allow them to cure for 1 week before eating. Properly sealed jars will keep in a cool, dark place for at least 6 months; refrigerate after opening.

WATERMELON RIND MOSTARDA

MAKES 2 PINTS

Pickled watermelon rind is a classic in the South, and rightfully so. It may seem weird, but it's really tasty. I wanted to take the concept a little further and develop some deeper flavors, so I looked to the Italians for inspiration and imported the time-honored mostarda technique. Mostarda is a condiment made with various fruits and a mustard-based syrup. This one is especially good served with wild game, and if you have some foie gras lying around, all you'll need to add is a piece of buttery toast for one hell of a snack.

One 3¾-pound piece watermelon with a thick rind

1 cup sugar

1 cup white vinegar

2 tablespoons fresh lemon juice

2 whole cloves

1 whole star anise

1 tablespoon prepared horseradish

1 teaspoon dry mustard

1 teaspoon white mustard seeds

1 teaspoon brown mustard seeds

1 small jalapeño pepper, seeded and cut into very small dice

Cut the watermelon flesh away from the rind; reserve the flesh for another use. Cut the green outer skin from the rind and discard. Cut enough of the white rind into ½-inch dice to measure 2 cups.

Combine the sugar, vinegar, lemon juice, cloves, and star anise in a medium stainless steel saucepan and bring to a boil over high heat, stirring to dissolve the sugar. Remove from the heat, add the watermelon rind, and stir to combine. Transfer to a glass or stainless steel container and cool for 1 hour, then cover tightly and refrigerate for 5 days.

Drain the rind, reserving the liquid. Combine ½ cup plus 2 tablespoons of the liquid, the rind, horseradish, dry mustard, mustard seeds, and jalapeño pepper in a medium stainless steel saucepan and bring to a simmer over medium heat. Remove from the heat.

Divide the mixture between two clean pint canning jars. Cool for 1 hour, then cover and refrigerate. Tightly covered, the mostarda will keep for up to 1 month in the refrigerator.

=== NOTE ===

To put up the mostarda for longer storage, follow the canning instructions on page 208.

DAVID SHIELDS

When I set out to really try to understand Lowcountry cooking, I knew that I had to go beyond the kitchen. I knew a lot about food, but I didn't necessarily know about what people call "foodways"—all of the connections between material culture, agriculture, economy, taste, and regional preferences that make up what we might call a cuisine. To understand a cuisine, the question to ask is not, "What do we cook?" Instead, it should be, "Who does the cooking, and why do they do it in this particular way?" You must understand more than simply the recipes.

Early on, I became close friends with Glenn Roberts and he often spoke of his friend David Shields, who was equally passionate about the restoration of Lowcountry cooking. David is a professor of Southern letters at the University of South Carolina. He's an expert on a bunch of esoteric things, from early film to opera, but he has dedicated much of his time to researching the plants and cooking of the Carolina Gold Rice kitchen. His book *The Golden Seed* details the history of our beloved rice, and he recently finished another book, called *Southern Provisions,* filled with all of the helpful information that he's shared with me over the last few years.

When I went searching for the essence of Lowcountry cooking, I found that there is a lot of misinformation out there. But David is all about the facts. He's an academic, and he bases his findings on sound research. When he tells me something, it's not some romantic story passed down through generations. Those often get bent along the way for the sake of good storytelling, or become

whitewashed when concerning the more difficult facets of Southern history.

David likes evidence and spends much of his time sifting through old seed catalogs, agricultural journals, newspaper clippings, and dusty antiquarian cookbooks from the nineteenth century. Thanks to his guidance, I am now addicted to researching these documents as well. David's studies have allowed me to develop a deep understanding of the cuisine of the Carolina Gold Rice kitchen—how it developed, what happened along the way, and where it's headed. If I have a question about a particular plant, or a dish, or even a period of time, I simply send David an e-mail. I receive a reply that often leaves me speechless.

My smartphone is full of historical documents that David has sent me. I often reference them on the spot when I'm cooking or having a conversation about Southern food. Simply put, my cuisine wouldn't be where it is today if I didn't have David as a primary resource.

David once said to me, "You can learn a lot more about a cuisine from agricultural literature than you ever can from a cookbook." What he meant was that in order to understand a cuisine, you have to understand the history of its agriculture and what the soil is capable of. So what makes a cuisine? After connecting with David, and tracing the Lowcountry to its beginnings, I think I can finally answer that question. In my humble opinion, a cuisine is based on two things: the ingredients and products that are available at the practitioner's fingertips combined with the cultural influences of the particular region.

PICKLED ELDERBERRIES

MAKES 1 QUART

The flavor of elderberries can be complex—bitter, sour, sweet, tannic. It all depends on when you harvest them. The berries are tiny and it takes forever to pick enough to amount to anything. Maybe that's part of the allure. The more work required in making something, the better it usually tastes.

Elderberries grow wild all over the place, especially alongside the road. Once you train your eye for them, you can spot the big flower heads a mile away. You might think eucalyptus is only used in cold remedies; think again. Here the eucalyptus adds a touch of that familiar menthol flavor.

1 cup white vinegar
½ cup water
1 tablespoon sugar
1 tablespoon yellow mustard seeds
10 fresh eucalyptus leaves (see Resources, page 326)
1 cup fresh elderberries
1 cup fresh lemon juice

Bring the vinegar, water, sugar, mustard seeds, and eucalyptus leaves to a boil in a stainless steel saucepan over medium-high heat, stirring to dissolve the sugar.

Combine the elderberries and lemon juice in a clean quart canning jar. Pour the hot pickling liquid over them. Cool to room temperature, then cover tightly and refrigerate for at least 1 month before eating to allow the elderberries to cure. Tightly covered, the elderberries will keep for up to 3 months in the refrigerator.

NOTE

To put up the elderberries for longer storage, follow the canning instructions on page 208.

HEIRLOOM TOMATO CONSERVE

MAKES 2 PINTS

This versatile little preserve is a great condiment to have in your pantry if for no other reason than to save you in a pinch when friends unexpectedly pop over. I crack open a jar and spoon it out over a hard sheep's-milk cheese. It cuts the salty bite of a great aged Pecorino with ease, but I've found that it's also fantastic over beautiful seared scallops. The unexpected flavors of yuzu and ginger make it unique.

4 large heirloom tomatoes (about 4 pounds)
1 cup dry white wine
1 cup white wine vinegar
1 cup sugar
2 medium shallots, cut into ⅛-inch dice
1 tablespoon grated peeled fresh ginger (use a Microplane)
1 cinnamon stick
1 whole star anise
1 tablespoon yuzu juice (see Resources, page 326)
Kosher salt

Bring a large pot of water to a boil. Make an ice bath in a bowl with equal parts ice and water. Submerge the tomatoes in the boiling water for 20 seconds. Remove and submerge them in the ice bath to cool them; do not leave them in there for longer than 5 minutes. Drain, then peel, halve, and seed the tomatoes and cut into ¼-inch dice.

Combine the wine, wine vinegar, sugar, shallots, ginger, cinnamon stick, and star anise in a medium stainless steel saucepan, bring to a boil over high heat, and boil until reduced by a third, about 10 minutes. Gently fold in the tomatoes and bring the mixture back to a boil.

Transfer the conserve to a glass or stainless steel container. Cool to room temperature, then add the yuzu juice. Season lightly with salt.

Divide the conserve between two clean pint canning jars. Cover and refrigerate. Tightly covered, the conserve will keep for up to 1 month in the refrigerator.

NASTURTIUM CAPERS

MAKES 1½ PINTS

Capers are traditionally made with the unopened seed pods of a Mediterranean flower bush, but nasturtium buds make a fine stand-in and have their own zesty kick. The best way to get nasturtium buds is to grow them yourself—you can find seeds and plants at your local garden store or on the Internet. Pick the buds that are soft and green, just before they open up; if you miss some, they'll give you beautiful little flowers that are edible as well.

This recipe pops up in many early American cookbooks. Nasturtiums were commonplace back then. Plant some and add this back to the pantry again.

2 cups nasturtium buds
3 cups water
4½ tablespoons kosher salt
1 cup white wine vinegar

Place the nasturtium buds in a glass or stainless steel bowl. Put 1 cup water in a small bowl, add 1½ tablespoons of the salt, and stir until it dissolves. Pour the water over the seed pods, cover, and allow them to sit at room temperature for 24 hours.

Drain the buds and repeat the process two times over the course of 3 more days, using 1 cup water and 1½ tablespoons salt each time.

Drain the buds and put them in 2 clean canning jars. Bring the vinegar to a simmer in a small saucepan and pour it over the nasturtium buds. Cool to room temperature, cover tightly, and allow to stand at room temperature for 1 week, then refrigerate. Tightly covered, the nasturtium buds will keep for up to 3 months in the refrigerator.

NOTE

To put up the capers for longer storage, follow the canning instructions on page 208.

PICKLED CABBAGE

MAKES 4 QUARTS

I have great memories of making this fermented cabbage with my grandmother. Each year when the cabbage was ready in the garden, we would chop it and put it into her ceramic crocks. The cabbage would sit in the basement and get fermented and delicious. Throughout the year, it made its way onto the dinner table as a side dish, kind of like the way kimchi is served alongside Korean food. Keep in mind that the longer you let this ferment, the funkier it gets. I like mine pretty funky, but you can adjust the timing to your tastes.

3 large heads green cabbage, with large outer leaves
1 cup plus 3 tablespoons kosher salt

Wash and dry the cabbage. Pull off the outer leaves. Stack them a few at a time, roll them into cylinders, and cut them into large ribbons. Core the heads and cut them into quarters, then thinly slice the cabbage.

Toss the cabbage and the salt in a large nonreactive container and mix well to combine. Weight the cabbage down with a plate to keep it submerged in the liquid that will be released.

Cover the container with a square of cheesecloth and tie it snugly. Let stand in a cool place for 1 week to cure, then refrigerate. Tightly covered, the cabbage will keep for up to 2 weeks in the refrigerator.

BUTTER-BEAN CHOWCHOW

MAKES 3 QUARTS

People say that the name "chowchow" comes from the French word for cabbage, *chou,* and that it was brought south by Acadians when they were expelled from Nova Scotia and made their way to what became Cajun country. Of course, a chow chow is also a breed of dog, so maybe we should just call this dish a relish. This version can be made with any Southern bean or pea you find fresh at the market. If you leave out the butter beans altogether, you will still have a great recipe for a simple, classic chowchow that can be modified as you see fit.

If you make this recipe for canning, process the full 3 quarts. But if you want to make a quick pickle only, you can halve the recipe, as it won't keep for very long in the refrigerator.

12 ounces butter beans, assorted varieties

6 cups apple cider vinegar

1½ cups packed light brown sugar

2 tablespoons salt

1 tablespoon yellow mustard seeds

1½ teaspoons celery seeds

1 tablespoon turmeric

1½ teaspoons crushed red pepper flakes

1 medium sweet onion (about 1 pound), cut into small dice

2 red bell peppers (about ¾ pound), cored, seeded, and cut into small dice

1 jalapeño pepper, seeded and minced

1 small head green cabbage (about 1½ pounds), cored and finely chopped

3 green tomatoes (about 1½ pounds), cored and cut into small dice

½ cup yellow mustard

Bring a large pot of heavily salted water to a boil over high heat. Add the butter beans and cook for 4 minutes. Drain the beans and spread them out on a baking sheet to cool.

Combine the vinegar, brown sugar, and salt in a medium stainless steel pot and bring to a boil over high heat, stirring to dissolve the sugar and salt. Reduce the heat to medium-high and cook the mixture for about 20 minutes, until reduced by half.

Add the mustard seeds, celery seeds, turmeric, and red pepper flakes and stir well. Add the onions, bell peppers, jalapeño pepper, cabbage, and tomatoes and cook over medium-high heat, stirring occasionally, until the cabbage is tender, about 15 minutes. Fold in the butter beans and yellow mustard and remove the pot from the heat.

Divide the chowchow among three clean quart canning jars. Cool to room temperature, then cover and refrigerate for at least 3 days before eating. Tightly covered, the chowchow will keep for up to 5 days in the refrigerator.

═══ NOTE ═══

To put up the chowchow for longer storage, follow the canning instructions on page 208.

BREAD-AND-BUTTER PICKLES

MAKES 3 QUARTS

Bread-and-butter pickles are a classic, and this is a pretty straightforward recipe to try if you're just starting out pickling. I took some old bread-and-butter pickle recipes, upped the spices, and cut back on the sugar. You can use this basic recipe to pickle all kinds of things. Green tomatoes are wonderful in place of the cucumbers, or try summer squash or zucchini.

2 pounds medium Kirby cucumbers

2 cups small dice sweet onion

¼ cup kosher salt

1 pound crushed ice

2 cups sugar

1 tablespoon yellow mustard seeds

½ teaspoon celery seeds

1 teaspoon turmeric

2 cups cider vinegar

1 fresh bay leaf, torn into pieces

3 whole cloves

½ jalapeño pepper, sliced paper-thin

Sterilize three quart canning jars, along with the rings and lids (see page 208).

Cut the blossom end off the cucumbers and slice the cucumbers into ⅛-inch-thick slices. To draw out the excess liquid and increase the crunch, layer the cucumber slices with the onion, salt, and ice in a glass or stainless steel container. Weight with a plate large enough to cover all the cucumbers and keep them submerged when the ice melts. Cover and refrigerate overnight.

Drain the cucumbers and onions, put them in a large stainless steel pot, and add the sugar, mustard seeds, celery seeds, turmeric, vinegar, bay leaf, cloves, and jalapeño pepper. Bring the mixture to a simmer over high heat, stirring to dissolve the sugar. Using a digital thermometer to check, make sure that the internal temperature of the cucumbers reaches 180°F and stays there for 90 seconds. Skim off any foam that rises to the top.

Divide the hot cucumbers among the sterilized jars. Ladle the remaining hot vinegar mixture over the cucumbers, leaving a ¼-inch headspace. Wipe the rims and threads clean. Place the lids and rings on the jars and tighten the rings. Cool the jars on a clean dish towel or a rack, not directly on the countertop.

Refrigerate for at least 1 week to let the pickles cure before eating. Unopened, the pickles will keep for up to 6 months in the refrigerator. Once opened, they will keep for up to 3 weeks longer in the refrigerator.

NOTE

To put up the pickles for longer storage, follow the canning instructions on page 208.

TOMATO JAM

MAKES 1 PINT

This recipe is an old standby for me. The method is very similar to making tomato paste. The idea is to cook down the tomatoes in a little vinegar with some spices until the natural flavors are concentrated and deeply transformed. I've tried just about every combination of vinegar and spices you could imagine and finally settled on a little nutmeg and Worcestershire. The jam is really good on fish and pan-roasted chicken, and it's a great way to use those excess or blemished tomatoes that come cheap at the height of summer.

3 large ripe heirloom tomatoes (about 3 pounds)

2 cups cider vinegar

1 cup packed light brown sugar

½ cup Worcestershire sauce

½ cup extra-virgin olive oil

Kosher salt

1 nutmeg

Bring a medium pot of water to a boil. Make an ice bath in a bowl with equal parts ice and water. Submerge the tomatoes in the boiling water for 20 seconds. Remove and submerge them in the ice bath to cool them; do not leave them in there for longer than 5 minutes. Drain, then peel, halve, seed, and chop the tomatoes.

Combine the vinegar and sugar in a stainless steel saucepan and bring to a boil over high heat, stirring to dissolve the sugar. Boil until reduced by half, about 7 minutes. Add the tomatoes and Worcestershire sauce and bring back to a boil, then reduce the heat to low and simmer the mixture, stirring frequently, until it is a dark brown color and very thick, about 20 minutes.

Transfer the jam to a blender, add the olive oil, and blend on high until smooth, about 2 minutes. Season with salt and 1 or 2 gratings of nutmeg (use a Microplane). Pour the jam into a clean pint canning jar, cover, let cool, and refrigerate.

Tightly covered, the jam will keep for up to 3 weeks in the refrigerator.

DILLED GREEN TOMATOES

MAKES 3 QUARTS

Dilled green tomatoes are great on their own, dunked in Bloody Marys, or coated in cornmeal to make fried pickled green tomatoes.

10 garlic cloves

3½ cups white vinegar

3½ cups water

1 cup sugar

¼ cup kosher salt

7 fresh bay leaves

1 tablespoon dill seeds

½ tablespoon celery seeds

½ tablespoon yellow mustard seeds

9 whole black peppercorns

3 large dill sprigs

3 tablespoons ⅛-inch-dice Vidalia onion

3½ pounds green heirloom tomatoes, sliced into ½-inch rounds

Sterilize three quart canning jars, along with the rings and lids (see page 208).

Cut 7 of the garlic cloves in half; leave the other 3 whole. Combine the vinegar, water, sugar, salt, the halved garlic cloves, and bay leaves in a medium stainless steel pot and bring to a boil over high heat, stirring to dissolve the sugar and salt. Turn off the heat but leave the pot on the stove while you fill the jars with the spices.

Put 1 teaspoon dill seeds, ½ teaspoon celery seeds, ½ teaspoon yellow mustard seeds, 3 peppercorns, 1 dill sprig, 1 tablespoon onion, and 1 whole garlic clove in each sterilized jar. Fill the jars with the green tomato rounds and cover with the hot pickling liquid, leaving a ½-inch headspace. Wipe the rims and threads clean. Put the lids and rings on the jars and finger-tighten the rings.

Process the jars in a boiling-water bath for 10 minutes according to the canning instructions on page 208.

It is important that the jars seal properly and a vacuum forms (see page 208). If any jars did not seal, you must store the tomatoes in the refrigerator; allow them to cure for 1 week before eating (see Note below for storing). Properly sealed jars will keep in a cool, dark place for up to 6 months; refrigerate after opening.

═══ NOTE ═══

This recipe also works as a quick pickle. Put the filled jars in the refrigerator, and the tomatoes will be ready in about 1 week; they will keep for 2 weeks after being opened.

PICKLED RAMPS

MAKES 3 QUARTS

Ramps are my favorite ingredient. In fact, I have a couple of them tattooed on my left arm. I go a little crazy when they come into season, which is a very short period. They really only grow wild in the mountains in the springtime, and that makes them hard to get. When our first delivery arrives, I eat nothing but ramps and cornbread for a couple of days.

Each year we buy close to a thousand pounds of ramps to preserve for later in the year. I have been working on this recipe for years, tweaking the spices, the sharpness, and the sweetness, and I am finally done messing with it. It tastes exactly the way I want it to.

2½ pounds ramps, cleaned and leaves and hairy root ends removed

¼ cup ½-inch-thick slices jalapeño peppers, seeds included

1¼ cups cider vinegar

1¼ cups water

1¼ cups rice wine vinegar

1¼ cups sugar

1½ teaspoons coriander seeds

¾ teaspoon fennel seeds

¾ teaspoon whole black peppercorns

3 whole cloves

1½ star anise

1 cinnamon stick

1 green cardamom pod, cracked

1 fresh bay leaf

Sterilize three quart canning jars, along with the rings and lids (see page 208).

Combine all of the ingredients in a large stainless steel pot and bring to a boil over high heat. Remove the pot from the heat.

Divide the ramp mixture among the canning jars. Wipe the rims and threads clean. Place the lids and rings on the jars and tighten the rings. Cool the jars on a clean dish towel or a rack, not directly on the countertop.

Refrigerate for 1 week before eating to allow the ramps to cure. Unopened, the ramps will keep for up to 4 months in the refrigerator. Once opened, they will keep for 3 weeks in the refrigerator.

The ramps can also be processed according to the hot-water-bath canning instructions on page 208. Process for 10 minutes. Properly sealed, the jars will keep in a cool, dark place for up to 6 months; refrigerate after opening.

PICKLED OKRA

Here's your chance to try one of the best pickled okra recipes I have ever tasted. It's one I learned when I was an extern at Peninsula Grill under chef Robert Carter, one of my mentors, and I've made this pickled okra every year since.

5 pounds medium okra, washed
7½ cups cider vinegar
3 cups water
7½ jalapeño peppers, thinly sliced into rounds, seeds included
7½ garlic cloves, thinly sliced
1½ cups kosher salt
¼ cup plus 2 tablespoons sugar
1½ tablespoons turmeric
1½ tablespoons yellow mustard seeds

Sterilize five quart canning jars, along with the rings and lids (see page 208).

Make a small slit under the cap of each okra pod so that the pickling liquid can enter the pods. Pack the okra tightly into the jars.

Combine the vinegar, water, jalapeño peppers, garlic, salt, sugar, turmeric, and mustard seeds in a large stainless steel pot and bring to a boil over high heat, stirring to dissolve the salt and sugar.

Ladle the mixture over the okra, leaving a ½-inch headspace. Wipe the rims and threads clean. Put the lids and rings on the jars and finger-tighten the rings.

Process the jars in a boiling-water bath for 15 minutes according to the canning instructions on page 208.

It is important that the jars seal properly and a vacuum forms (see page 208). If any jars did not seal, you must store the okra in the refrigerator; allow them to cure for 1 week before eating (see Note below for storing). Properly sealed jars will keep in a cool, dark place for up to 6 months; refrigerate after opening.

NOTE

This recipe also works as a quick pickle. Halve the recipe, and store the jars in the refrigerator. The okra will be ready in about 1 week; it will keep for up to 3 weeks after opening.

CURED EGG YOLKS

I bet cured egg yolks were invented because someone had more eggs on hand than they knew what to do with. It's a simple idea: use salt and sugar to cure the yolks. I shave cured yolks over salads or vegetables and anywhere you might use Parmesan cheese. Use them to garnish egg dishes too.

1¾ cups plus 2 tablespoons kosher salt
½ cup plus 2 tablespoons sugar
12 large eggs, separated, whites reserved for another use

EQUIPMENT
Butcher's twine

Mix the salt and sugar together. Spread half of the mixture in the bottom of a 9-by-12-inch baking dish. Make 12 small indentations to hold the yolks. Carefully place a yolk in each indentation and cover them with the remaining salt and sugar mixture. Cover and refrigerate for 1 week.

After 7 days, brush off as much salt and sugar as possible from the yolks. Lay a double thickness of cheesecloth on a work surface. Line the yolks up on it side by side, leaving space between them. Roll up the yolks in the cheesecloth. Using butcher's twine, knot the cheesecloth between the yolks (as if you were tying off sausage links).

Let the roll hang in a pantry or other enclosed cool area for 1 week. After that, the yolks can be held in an airtight container for up to 2 months.

HOMEMADE STEAK SAUCE

MAKES 1¼ PINTS

Most chefs wouldn't admit this, but I really like steak sauce. If you think about it, bottled steak sauces are actually pretty complex, packed with acidity, sourness, and enough meaty umami to satisfy any foodie out there. My recipe is a mash-up of sorts, inspired by my favorite bottled steak sauces, and I think it's damn tasty.

4 whole cloves
1 whole star anise
1½ cups ketchup
1 cup Worcestershire sauce
½ cup water
½ cup fresh orange juice
½ cup white vinegar
¼ cup soy sauce
2 tablespoons fish sauce
¼ cup packed light brown sugar
2 tablespoons Dijon mustard
½ cup small dice sweet onion
½ cup golden raisins
2 garlic cloves, thinly sliced
2 tablespoons grated fresh ginger (use a Microplane)
Grated zest of 1 orange (use a Microplane)

EQUIPMENT
Butcher's twine

Wrap the cloves and star anise in a square of cheesecloth and tie it with butcher's twine.

Combine all the remaining ingredients and the spice sachet in a large stainless steel saucepan and bring to a simmer over high heat, stirring to dissolve the sugar. Reduce the heat to medium-high and cook the mixture, stirring frequently to prevent scorching, until it is reduced by half, about 35 minutes.

Remove the spice sachet, transfer the mixture to a blender, and blend on high until smooth, about 3 minutes. Spoon the steak sauce into two clean pint canning jars. Cool to room temperature, then cover and refrigerate. Tightly covered, the sauce will keep for up to 2 weeks in the refrigerator.

SHAD ROE BOTTARGA

MAKES 6

I call bottarga the country ham of the fish world. Use it anywhere you'd use country ham or to impart a robust tinge of the sea. Grate it over salads (in place of anchovies), baked oysters, or even pasta. During the spring shad run, the fish are caught in multitudes, and that equals a lot of roe, which is highly perishable. The locals like to fry up the roe sacs in butter and serve them over grits for breakfast, but we always purchase a big batch to salt down into this Italian-style condiment.

16 pounds kosher salt
6 shad roe sacs
Dry white wine
Extra-virgin olive oil

Place 4 cups of the salt in a pan big enough to hold the roe sacs without touching one another. Place the roe sacs on the salt. Cover the sacs completely with another 4 cups salt.

Place the pan in a cool, dark, dry place. The temperature should be between 50°F and 70°F. Let stand for 3 days.

Rinse the salt off the roe sacs and pat them dry. Wipe out the pan and repeat the curing process 4 more times over the course of 2 weeks, using 8 cups salt each time.

The roe sacs should be hard as a rock. If they aren't, repeat the process until they are hard.

Rinse the roe sacs with white wine and dry them. Store the roe sacs sealed airtight in a plastic container in a cool, dark place for up to 6 months.

When you are ready to use one of the roe sacs, remove the outer membrane and grate the roe with a Microplane. After using it, coat the remainder in a light film of olive oil and store it in another airtight container.

CRISPY PIG'S-HEAD PANCETTA

MAKES ABOUT
5½ POUNDS

These crispy little "pig chips" are one of my favorite ways to introduce people to charcuterie. Give a squeamish man some souse or headcheese, and he's liable to run for the hills. But everyone likes a little crispy bacon, because pork fat is delicious, and some of the most delicious parts of the pig can be found in the head. You'll need a digital scale for this recipe.

The pancetta makes a lot, but it freezes well. You'll need to order the pig's head from a specialty or ethnic butcher.

132 grams kosher salt

26 grams sucrose (see Note)

14 grams TCM (see Note, page 150, and Resources, page 326)

1 gram freshly grated nutmeg (use a Microplane)

1 deboned pig's head (about 7½ pounds; have your butcher debone it)

9.5 grams freshly ground white pepper

5.1 grams ground mace

5.6 grams grated garlic (use a Microplane)

7.4 grams chopped rosemary

8.6 grams chopped thyme

EQUIPMENT

Digital scale

Butcher's twine

To make the cure, combine the salt, sucrose, TCM, and nutmeg in a small bowl. Divide the mixture in half.

Rub the inside of the head with half of the mix. Cover the remaining mix tightly and reserve. Fold the meat side of the head together cheek to cheek. Place it on a rack set on a rimmed baking sheet. To compress the head, place several heavy baking sheets on top of it to weight it down. Refrigerate it, with the weight on it, for 4 days.

After the 4 days, remove the head from the refrigerator. Add the pepper, mace, garlic, rosemary, and thyme to the reserved cure. Unfold the head and rub the meat with this mix. Refold the head, place the baking sheets on it again, and refrigerate for another 4 days.

Remove the head from the refrigerator and wipe off any excess cure on the outside with a damp towel. Tie up the head with butcher's twine and hang it to cure for 8 weeks in the refrigerator; the temperature should be between 34°F and 40°F. Place a pan filled with kosher salt underneath it to catch any drippings (and absorb odors). The head will lose about 30 percent of its weight during this time.

To make the crispy pancetta, preheat the oven to 350°F.

Shave thin slices from the head, place them on a rimmed baking sheet lined with a silicone mat, and roast them for 10 minutes, or until crisp and browned.

=== **NOTE** ===

Sucrose adds flavor to the meat and counteracts the saltiness. It's used for curing because it will be dispersed more evenly than granulated sugar.

Shave off only as much pancetta as you plan to eat within a day. Place the remainder of the whole head in a paper-towel-lined container, cover, and refrigerate for up to 1 month, or freeze, tightly covered, for up to 3 months.

SMOKED BACON
FOR BEGINNERS

MAKES ABOUT
6½ POUNDS

Here's an easy way to make your own bacon at home. If you are limited on space, just buy a small piece of pork belly, scale down the proportions of the salt rub relative to the weight of your meat, and give it a try. You'll soon want to graduate to bigger pieces, though, and eventually whole slabs of belly. You never know, but you might turn into the next Allan Benton (although I don't want to get your hopes up).

5 cups kosher salt

1 cup packed light brown sugar

¼ cup crushed red pepper flakes

1 teaspoon TCM (see Note, page 150, and Resources, page 326)

1 tablespoon coarsely ground black pepper

1 pork belly (about 12 pounds)

EQUIPMENT

3 pounds hickory wood chips, soaked in cold water for a minimum of 30 minutes but preferably overnight

Butcher's twine

To make the cure, combine the salt, sugar, red pepper flakes, TCM, and black pepper in a bowl.

Remove the skin from the pork belly. Reserve it to make pork rinds (see page 254).

Rub the belly thoroughly with the cure. Wrap the belly tightly in plastic wrap and place it on a rimmed baking sheet. To compress the belly, place several heavy baking sheets on top of it to weight it down. Refrigerate it with the weight on it for 1 week, turning it over every day.

After a week, rinse the belly and pat it dry. Heat your smoker to 250°F.

Hot-smoke the belly over hickory chips until it reaches an internal temperature of 150°F, about 4 hours (see sidebar, page 160). Cool the belly to room temperature.

Tie up the belly with butcher's twine, or use a bacon hook, and hang it in the refrigerator for 30 days. Make sure the temperature stays below 40°F, and place a pan filled with kosher salt underneath it to catch any drippings (and absorb odors). The bacon will experience moisture loss and the texture will become firmer. After 30 days, tightly wrap the bacon in plastic wrap and refrigerate; it will keep for up to 6 months in the refrigerator.

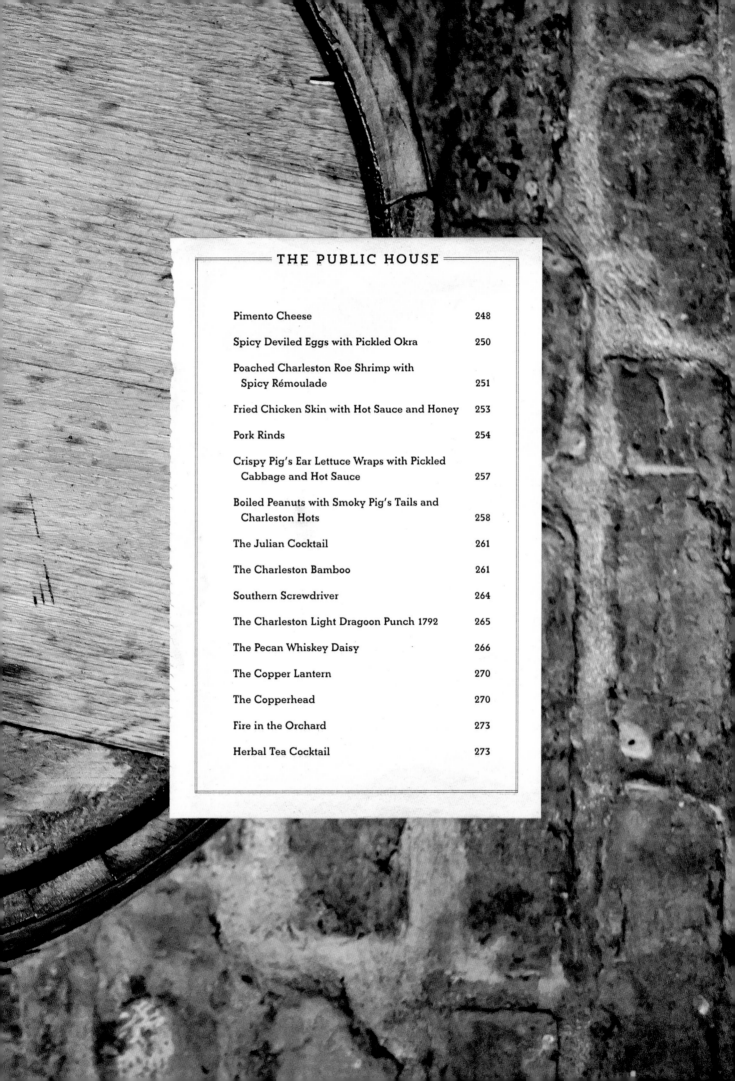

THE PUBLIC HOUSE

SOUTHERN SPIRIT

I n the South, we sure do enjoy a good party. Cocktails and spirits have always been a way for us to make people feel comfortable when they pay us a visit. Cocktails are a way of life for many, oftentimes the reward at the end of a long day of work or the beginning of a celebration or a ball game or a wedding—we hardly need an excuse to raise a glass. The South is full of that celebratory spirit. A drink and a snack will make you relax, let you slow down and, most of all, have a good time.

A public house is a traditional gathering place for food and drink. McCrady's Tavern was a public house in the years following the American Revolution and throughout the first half of the nineteenth century. Edward McCrady was a war hero who purchased the tavern and even hosted George Washington there. Back then, bars were a place where people gathered not just to socialize, but to air complaints, organize political movements, and wage wars, often fortified by alcoholic beverages.

Perhaps the beverage the South is most associated with is moonshine, a spirit that bore its own sport: the first NASCAR races were, after all, "stocked" with moonshiners' souped-up cars. Those early races were a testament to not only the industrious ingenuity of Southerners, but also their willingness to thwart authority when it threatened their perceptions of freedom.

Elijah Craig, a Virginian, had the brilliant idea to age moonshine in charred oak casks. In 1789, he founded a distillery, and American bourbon was born. A lot of people don't really understand what makes bourbon bourbon. Whiskey is a spirit distilled from grain at below 160 proof and aged in oak barrels. There are many different whiskeys: Scotch, Canadian, blended, Japanese, and American. But to be called bourbon whiskey, the whiskey has to meet several U.S. regulations established by Congress in 1964.

Bourbon has to be made in the United States. It has to be distilled from a mash of at least 51 percent corn but below 80 percent corn and below 160 proof. It has to be aged in charred new oak barrels. And it cannot have any added color or flavoring.

Bourbon derives its name from Bourbon County, Virginia, which was named in honor of the French royal house of Bourbon. Corn, a crop indigenous to the region and important to the culture, is made into booze. Then, to make it something special, it's aged, just like a country ham. It can age for anywhere from a few months to years, which means making bourbon takes time and patience—the two most important ingredients.

This chapter is an introduction to some of the cocktails and snacks that you would find at my home and at the bars at my restaurants. I like to drink bourbon straight, but I also like to make bourbon cocktails, so you'll find recipes for some of those cocktails, including The Julian Cocktail (page 261) and The Pecan Whiskey Daisy (page 266). You'll also find instructions for setting up a cocktail station for parties (see page 262).

If you walk into the bar at Husk, you'll see a wall of bourbon. It's a spirit that has endured the test of time. I think a craving for whiskey is in every Southerner's DNA. Babies get bourbon rubbed on their gums. Sick kids get a hot toddy before bedtime. Today bourbon is growing in popularity, and that's a major source of pride in the South that we celebrate.

PIMENTO CHEESE

MAKES 2½ TO 3 CUPS

I've seen people almost get into fistfights over who has a better pimento cheese recipe. Southerners don't mess around when it comes to their cherished "pâté de Sud." We slather the stuff on everything from celery stalks to saltine crackers, and some people won't even consider eating a hamburger without a half-inch layer of pimento cheese in the stack.

Everyone has his or her own way of making pimento cheese, but the biggest debate always revolves around what kind of mayo is used. I prefer Duke's; it happens to be my favorite. But you can use your favorite brand—that's what making a signature pimento cheese is all about. Of course this is best made with pimento peppers you roast yourself, but if you can't get the fresh peppers, substitute 12 ounces jarred whole pimentos, drained and diced (don't use jarred chopped pimentos—they have no flavor).

3 large pimento peppers (about 12 ounces)

4 ounces cream cheese, at room temperature

½ cup mayonnaise, preferably Duke's (see Resources, page 326)

½ teaspoon Husk Hot Sauce (page 238)

½ teaspoon kosher salt

¼ teaspoon sugar

⅛ teaspoon cayenne pepper

⅛ teaspoon freshly ground white pepper

⅛ teaspoon smoked paprika (see Resources, page 326)

¼ cup Pickled Ramps (page 233), chopped, plus ½ cup of the brine

1 pound sharp cheddar cheese, grated on the large holes of a box grater

Roast the peppers over an open flame on a gas stovetop, one pepper at a time, on the prongs of a carving fork. Or place on a baking sheet and roast under a hot broiler. In either case, turn the peppers to blister all sides. Then transfer the peppers to a bowl and cover the bowl with plastic wrap. Set aside to let the peppers steam until cool enough to handle.

Carefully peel the blackened skin off each pepper. Cut the peppers lengthwise in half, open out flat on a cutting board, and carefully scrape away all the seeds and membrane. Dice the peppers.

Put the cream cheese in a medium bowl and beat it with a wooden spoon until softened. Add the mayonnaise and mix well. Add the hot sauce, salt, sugar, cayenne pepper, white pepper, and smoked paprika and stir to blend. Add the ramps, ramp brine, and cheddar cheese and stir again. Fold in the diced pimentos.

Cover and refrigerate until ready to serve. Tightly covered, the pimento cheese will keep for up to 3 days in the refrigerator.

NOTE

For creamer pimento cheese, combine all of the ingredients in the bowl of a stand mixer fitted with the paddle attachment and beat on medium speed for 2 minutes.

SPICY
DEVILED EGGS WITH
PICKLED OKRA

MAKES 12
DEVILED EGGS

My take on deviled eggs is pretty straightforward, as deviled eggs should be. The most important thing is to buy good eggs. That means buying them at your farmers' market, or even directly from a farm. You'd be surprised how much work goes into getting eggs ready for sale at a small-scale farm. Each egg is usually inspected and cleaned by hand, and that makes a big difference in quality. I give my deviled eggs a splash of pickled okra juice along with a healthy dose of good mustard and as much hot sauce as my guests can stand.

6 eggs

1 tablespoon mayonnaise, preferably Duke's (see Resources, page 326)

1 tablespoon yellow mustard

1½ teaspoons brine from Pickled Okra (page 234), plus sliced rings of okra

2½ tablespoons small dice Bread-and-Butter Pickles (page 228)

2½ tablespoons small dice Pickled Ramps (page 233)

Husk Hot Sauce (page 238)

Smoked paprika (see Resources, page 326)

Using a sewing needle or pushpin, pierce a hole in the shell at the wide end of each egg. Put the eggs in a large saucepan and cover them with room-temperature water. Bring the water to a boil over medium-high heat and boil the eggs for 2 minutes. Remove the saucepan from the stove, cover it, and let the eggs remain in the water for 10 minutes.

Carefully drain the eggs in a colander in the sink, then peel the eggs under cold running water.

Cut the eggs crosswise in half. Remove the yolks and put them in a bowl. Add the mayonnaise and mustard and mash the yolks well, then whisk until smooth. Add the pickled okra brine, pickles, and pickled ramps and stir to combine. Add hot sauce to taste. Transfer to a pastry bag fitted with a star tip or a zip-top plastic bag (snip off one corner of the bag).

Cut a small slice off the bottoms of the egg whites so they will stand upright and line them up on a serving dish.

Pipe the yolk mixture into the whites. Garnish with pickled okra and smoked paprika and serve.

NOTE

The deviled eggs can be made up to a day ahead: once you remove the yolks from the whites, turn the whites upside down on a tray covered with paper towels, cover with plastic wrap, and refrigerate. Make the filling from the yolks, cover, and refrigerate. Thirty minutes before you are ready to serve, pipe the filling into the eggs and let them come to room temperature.

POACHED CHARLESTON ROE SHRIMP WITH SPICY RÉMOULADE

SERVES 6 TO 8

When the shrimp in Charleston are carrying roe, I get very excited. I'll buy several pounds of them in early June and just sit at home and gorge myself. The roe—really tiny eggs (think built-in shrimp caviar)—makes the shrimp especially sweet and the flesh meltingly soft. This rémoulade is a nice dipping sauce to have on the table, but I'm just as happy with a big pile of these poached beauties and a couple of lemon wedges.

4 cups Vegetable Stock (page 316)

1 cup dry white wine

1 tablespoon yellow mustard seeds

1 tablespoon whole black peppercorns

1 teaspoon celery seeds

1 fresh bay leaf

2 lemons, cut in half

¼ cup Husk BBQ Rub (page 311)

2 pounds roe shrimp or large fresh wild American shrimp (16–20 count) in the shell

Juice of 1 lemon

Spicy Rémoulade (recipe follows)

Combine the vegetable stock, wine, mustard seeds, peppercorns, celery seeds, bay leaf, lemons, and 2 tablespoons of the barbeque rub in a medium stainless steel pot and bring to a boil over high heat. Reduce the heat and simmer for 30 minutes.

Remove the pot from the heat, place the shrimp in the pot, and cover the pot. Allow the shrimp to poach in the stock, off the heat, for 7 minutes. They should be pink and just beginning to curl. Remove the shrimp. Strain the flavorful stock and freeze to use in soups or sauces.

Toss the shrimp with the remaining 2 tablespoons rub and the lemon juice. Serve the shrimp warm, peel-and-eat style, in a large bowl with a side of rémoulade.

SPICY RÉMOULADE

MAKES 4 CUPS

2 cups mayonnaise, preferably Duke's (see Resources, page 326)

½ cup plus 2 tablespoons Creole mustard

½ cup ketchup

¼ cup minced celery

¾ cup minced shallots

¾ cup chopped scallions

1½ teaspoons minced garlic

2 tablespoons chopped capers

2 tablespoons brine from Pickled Okra (page 234)

1 tablespoon plus ¾ teaspoon fresh lemon juice

1 tablespoon prepared horseradish

1 tablespoon Husk Hot Sauce (page 238)

1 teaspoon smoked paprika (see Resources, page 326)

½ teaspoon cayenne pepper

Combine all the ingredients in a bowl or a refrigerator container with a lid and mix well. Cover and refrigerate until chilled. Tightly covered, the sauce will keep for up to 5 days in the refrigerator.

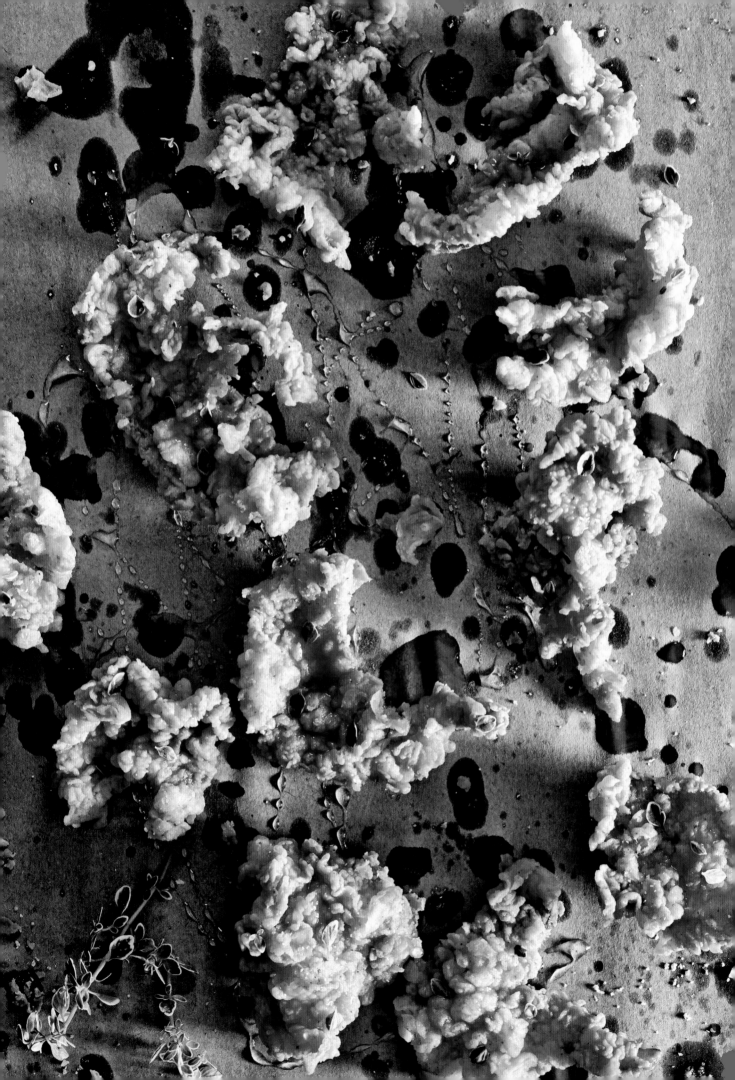

FRIED CHICKEN SKIN
WITH HOT SAUCE AND HONEY

SERVES 4

Almost everyone agrees that the guiltiest pleasure when eating fried chicken is the skin—don't act like it's not. I came up with this idea one day when I was eating some fried chicken at my favorite soul food restaurant in Charleston. It's a great little place called Martha Lou's Kitchen, and I eat there too frequently. Ms. Martha and her daughter sling out some true Southern cookin'—honest cooking that makes you feel good. That's why they call it soul food.

I always take my young cooks to taste Ms. Martha's food. Some people swear by the chitlins or the liver pudding she fries up for breakfast, but I almost always order the fried chicken. On one of these visits, I looked up at my chefs after a first bite of hot crispy chicken and said, "What if we just served the skin?" The table went silent, and I got a lot of crazy looks, but I thought, "Damn, people are going to love it!" I was the only one who thought it was a good idea.

I kind of forgot about it for a while, and then I came across a stack of boneless, skinless chicken breasts in a grocery store. It got me thinking about where all that skin went. So I called up an all-natural chicken operation in North Carolina and asked. They thought I was just as crazy as the chefs did, but the rest is history. Now we always have these fried chicken skins on our menu at Husk Bar; if we don't, our patrons complain.

Ask a good butcher to save some chicken skin for you.

2 cups ¼-inch-wide strips fresh chicken skin

2 cups whole-milk buttermilk

1 teaspoon cayenne pepper

1 teaspoon kosher salt

4 cups canola oil

4 cups all-purpose flour

1 teaspoon freshly ground black pepper

1 teaspoon garlic powder

1 teaspoon onion powder

1 teaspoon smoked paprika (see Resources, page 326)

Dark amber honey (see Resources, page 326)

Husk Hot Sauce (page 238)

Lemon thyme for garnish

Preheat the oven to 300°F.

With the back of a knife, scrape away all the visible fat from the chicken skin. The cleaner you get the skin at this point, the crisper it will be when it is fried. (You may want to refrigerate the fat and save it to use for flavor when cooking fried chicken or for rubbing on a chicken before you roast it.)

Combine the buttermilk, cayenne, and salt in a baking dish, add the skin, and gently toss. Bake the chicken skin in the buttermilk for 1 hour, or until tender. Remove the baking dish from the oven and leave the skin in the buttermilk until cool enough to handle.

Remove the skin from the buttermilk, shake off any excess, and lay it out on a wire rack set over a baking sheet to drain and cool completely, about 5 minutes.

Meanwhile, heat the canola oil to 350°F in a deep fryer or a large cast-iron skillet over medium heat. Combine the flour, black pepper, garlic powder, onion powder, and paprika in a baking dish.

Toss the skin in the flour mixture, coating it liberally. Working in batches, fry the skin for about 5 minutes, until golden brown and crisp. Drain the skin for a minute or two on wire racks covered with paper towels.

Serve the chicken skin family-style on a platter with a drizzle of honey, a splash of hot sauce, and a sprinkle of lemon thyme.

PORK RINDS

MAKES 24 SQUARES

I'll admit it right up front: this recipe is difficult, and it takes a lot of time, but it makes the best pork rinds I've ever tasted. We serve these often at Husk, but now people can make their own if they're willing to put in a little elbow grease! (Order the pork skin from a specialty or ethnic butcher.) You can serve these rinds for snacking or use them as a garnish for just about any dish when you want to add a porky crunch.

Note that the pork skin must be cooked for 24 hours and then dried for another 24 hours.

One 12-inch square pork skin
Peanut oil for deep-frying

EQUIPMENT
Immersion circulator
Vacuum sealer
Dehydrator

Preheat the water bath in an immersion circulator to 83°C (see sidebar, page 118).

Place the skin on a cutting board fat side up. Using a thin flexible knife, slice off as much fat as possible. The more fat you remove, the lighter the pork rinds will be. When you have removed all the fat, turn the knife over and scrape the skin with the back of the knife. You should be able to see through the skin. Discard the fat.

Place the skin in a vacuum bag and seal it on the highest setting. Cook the skin in the water bath at 83°C until it is soft, about 24 hours.

When the skin is cooked, make an ice bath in a bowl with equal parts ice and water and put the bag in the ice bath until the skin is completely chilled. Remove it from the bag and scrape the fat side of the skin one more time to make sure that all of the fat has been removed.

Cut the skin into twenty-four 1-inch squares. Place the pieces in a dehydrator set at 160°F and dehydrate them for 24 hours. They should snap when you try to break a piece off.

Tightly covered, the dehydrated pieces can be stored in a plastic container at room temperature for up to 1 week. Do not refrigerate them.

When ready to serve, pour 4 inches of oil into a deep pot and heat to 350°F over medium heat. Fry the pork rinds, in batches, for about 1 minute, until they are fully puffed. Drain for a minute or two on wire racks covered with paper towels, then serve with some hot sauce on the side.

CRISPY PIG'S EAR LETTUCE WRAPS

WITH PICKLED CABBAGE AND HOT SAUCE

MAKES 20 LETTUCE WRAPS

This is one of the very few dishes at Husk that has become a mainstay on the menu. I love it when I am walking down the street and overhear complete strangers talking about how much they love our crispy pig's ears. We came up with the technique of using a pressure cooker to soften the pig's ears before frying them. The result is crispy and wonderful, with a great pork flavor. Another great use for these ears is as a garnish on a plate of shrimp and grits.

PIG'S EARS

7 fresh pig's ears (see Note)

3 cups Pork Stock (page 319) or Chicken Stock (page 318)

2 quarts canola oil or a combination of 5½ cups peanut oil and 2½ cups Rendered Fresh Lard (page 316)

SAUCE

1 cup Husk Hot Sauce (page 238)

1 tablespoon soy sauce, preferably Bourbon Barrel Foods Bluegrass Soy Sauce (see Resources, page 326)

8 tablespoons (1 stick) unsalted butter, diced and kept cold

WRAPS

2 heads Bibb lettuce, separated into leaves (you need 20 leaves for the recipe), washed and patted completely dry

2 cups Pickled Cabbage (page 224)

2 tablespoons Anson Mills Antebellum Benne Seeds (see Resources, page 326), lightly toasted

Charleston coriander (optional)

EQUIPMENT

Pressure cooker

FOR THE PIG'S EARS: Place the ears and stock in a pressure cooker. Lock on the lid, bring the cooker up to high pressure, and cook for 90 minutes.

Line a baking sheet with paper towels. Carefully release the steam from the pressure cooker. Check the ears: they should be very soft, and you should be able to squeeze all the way through the cartilage. Carefully remove the ears, preferably using a slotted spatula. Lay the ears out in a single layer on the lined baking sheet and refrigerate until cool. Freeze the stock for another use.

When the ears are cool, cut them into thin strips about ¼ inch wide. *(This can be done up to 1 day ahead; keep the strips covered in the refrigerator and bring to room temperature before frying.)*

TO FRY THE EARS: Heat the canola oil to 350°F in a fryer or large cast-iron skillet. The oil will splatter when the strips hit the hot oil, so have a splatter guard or a lid at hand.

MEANWHILE, MAKE THE SAUCE: Mix the hot sauce and soy sauce in a small saucepan and heat over medium heat until warm, about 2 minutes. Remove from the heat and whisk in the butter a little at a time until emulsified.

Working in batches, fry the pig's ears until golden brown and crisp, 1½ to 2 minutes. Drain the strips for a minute or two on wire racks covered with paper towels.

TO COMPLETE: Arrange the lettuce leaves on a large platter. Toss the crispy ears in the hot sauce mixture and place them on the lettuce leaves. Top with the cabbage. Garnish with the benne seeds and Charleston coriander, if desired. Serve at once.

══ NOTE ══

Unless you live in the Deep South, where traditional grocers stock them regularly, pig's ears can be hard to find. They are sometimes available in markets catering to ethnic populations, especially Asian groceries, or you can find them online. Many farms will now ship directly to customers, and an Internet search will quickly find you someone with a pile of ears ready for the pot.

BOILED PEANUTS
WITH SMOKY
PIG'S TAILS AND
CHARLESTON HOTS

MAKES 4 CUPS

I am always surprised at how many people have never eaten boiled peanuts. Throughout the South, we have lots of little roadside stands that sell them every day of the week. In Charleston, there's a great one called Timbo's. It's a giant Airstream trailer set up on blocks and painted a bright orange, and I just can't help myself. I love having a Crock-Pot set on warm full of boiled peanuts at my house.

The boiled peanuts at the stands are pretty darn tasty, but this recipe takes it over the top by adding smoked pig's tails to the mix. If you have never enjoyed the pleasures of eating boiled peanuts at a baseball game or in a blind at a fall dove shoot, these will turn you on to one of the South's finest pleasures. Just make sure that you start with green peanuts—fresh peanuts that have not been cooked or dried. The recipe won't work any other way.

Note that the peanuts must be refrigerated overnight in the brine.

2½ pounds green peanuts (see Resources, page 326)

2 smoked pig's tails

2½ Charleston Hots (peppers; see Resources page 326), seeds removed, and sliced into thin rounds, or 2½ fresh cayenne peppers

¼ cup kosher salt

2 tablespoons chili powder

2¼ teaspoons light brown sugar

2¼ teaspoons onion powder

1½ teaspoons ground cumin

1 teaspoon dried oregano

¾ teaspoon paprika

¾ teaspoon garlic powder

¼ teaspoon cayenne pepper

⅛ teaspoon dry mustard

EQUIPMENT

Slow cooker

Combine all of the ingredients in a slow cooker and add water to cover. Cook for about 8 hours on high, replenishing the water as needed to keep the ingredients covered. The peanuts will float to the top when they are done and should be very tender on the inside.

Pour the peanuts and brine into a heatproof container and let cool to room temperature, about 30 minutes.

Pick the meat from the pig's tails, discard the bones, and add the meat to the peanuts. Cover the peanuts and brine and refrigerate overnight. *(Tightly covered, the peanuts can keep in the brine for up to 2 days in the refrigerator.)*

To serve, reheat the peanuts in the brine in a medium saucepan over low heat. Serve warm.

JULIAN VAN WINKLE

Bourbon's my thing—there's no denying it. I have a beat-up old pewter flask that I've traveled the world with. Each shared nip is my way of showing the world how special the South is. I have been responsible for converting a lot of people into nutcase bourbon lovers. The process of bourbon making symbolizes much of what I find important about Southern culture. Like the country ham that it often accompanies, bourbon is derived from a few simple, good-quality ingredients and depends on patience and passion and hard work to make it the legendary product that today is known worldwide as an emblem of the South.

My friend Julian Van Winkle III is also a bourbon man. He's a guy who struggled in the 1980s, trying to keep the family bourbon business afloat. But today, if you know anything about bourbon, you've probably heard about Julian and Old Rip Van Winkle bourbon. The Van Winkles of Louisville, Kentucky, are third-generation bourbon producers. I consider them family. I once conducted a blind tasting of bourbons at Tennessee's famed Blackberry Farm resort with Julian in attendance. Although I nailed the tasting and picked out the fifteen-year-old Pappy Van Winkle from a selection of five decanters, Julian had told me beforehand that he'd disown me if I failed. I was about as nervous as I have ever been, but the truth is that the task was easy.

I covet Van Winkle bourbon for its distinctiveness. Small-batch production and attention to detail render a product with deeply wrought flavor. A Van Winkle bourbon seems almost to have squeezed something extra from the corn; it has a mysterious complexity that is uniquely its own. It's a perfect example of old-fashioned American craftsmanship. People with a product like that could easily shift the focus of their business to extract much more profit, but then it wouldn't be Van Winkle anymore. The Van Winkle brand is built upon the tradition, hard work, and sacrifice that I'm convinced are the secret ingredients of not just Van Winkle bourbon, but of all traditional Southern culinary practices.

The Van Winkle family motto is, "At a profit if we can, at a loss if we must, but always fine bourbon." These are words that Julian lives by, and cuisine and spirits in America would be better off if everyone thought this way. Julian embodies something about the South that often gets missed: he perseveres. There were times when bourbon wasn't all that popular, and people weren't willing to mortgage a house to get a case of twenty-year-old Pappy, but Julian never stopped tending to what his family had built. He kept doing what his people always had done with the faith that quality would eventually trump the quick profit. He was right.

THE JULIAN COCKTAIL

MAKES 1 DRINK

This is a version of a drink that Julian Van Winkle made for me in his home one evening. He said it was the way his grandpa liked to mix a drink; it's a cocktail very similar to a classic old-fashioned.

The ritual of making the drink is the fun part. People love watching us make these behind the bar at Husk, and it's essential that you use a very strong (107-proof) bourbon. At first sip, the cocktail is quite strong, but pay attention to the flavor of the bourbon and try to pick out different tasting notes. As the ice melts in the drink, the proof drops and the flavor of the bourbon will change along with it. Each sip will be different. Drink it slowly. It takes a long time to make this perfect cocktail, and so it should take a long time to drink it.

1 orange wedge, unpeeled
1 raw sugar cube
7 dashes Angostura bitters
7 dashes orange bitters, preferably Regan's
2½ ounces 107-proof bourbon, preferably Pappy Van Winkle
Ice cube or sphere

Trim the tips off the ends of the orange wedge to cut down some of the bitterness and place the wedge in a large cocktail shaker. Cover the shaker with a bar napkin. Put the sugar cube in the center of the napkin and carefully coat the sugar cube with the bitters; the napkin will catch the runoff. Drop the soaked cube into the shaker and discard the napkin.

Using a bar spoon or a muddler, muddle the sugar with the flesh of the orange wedge; be careful not to push into the pith, as that would make the cocktail bitter. The result should be a thick syrup. Add 1 ounce of the bourbon and muddle for 3 minutes. Add a big ice cube or sphere. Add another ounce of bourbon and stir for 30 seconds. Pour into a rocks glass.

Top off with the remaining ½ ounce of bourbon and stir for 5 seconds. Serve immediately.

THE CHARLESTON BAMBOO

MAKES 1 DRINK

The classic bamboo cocktail was invented by Louis Eppinger at the Grand Hotel in Yokohama, Japan, in the late nineteenth century. Louis was a student of America's most famed and respected mixologist, Jerry Thomas. To give the drink a Charleston flavor, we substitute Madeira for the sherry. Madeira has a long history in Charleston, especially at McCrady's, where it once constituted the basis of many a wealthy gentleman's local drinking habit. Madeira is a fortified wine, and before the advent of refrigerated transport, fortified wines were the only ones that could make the trip across the Atlantic without losing their fine character. In fact, with time, Madeira would actually get better—so much so that wealthy Americans would buy barrels of the stuff and pay to have it ride in the belly of a ship halfway across the world and back. The older it got, sloshing to and fro all the way to the equator and back, the better it tasted upon its return.

My friend Mannie Burke owns the Rare Wine Co., which is the best source for Madeira in the world—he stocks barrels dating back to the eighteenth century—and it's where I get the spirits to make this cocktail. When peaches are in season, we make our own peach bitters and garnish the cocktail with a couple of slices of the ripe fruit.

Chipped ice
2 ounces dry Madeira, preferably Charleston Madeira
  (see Resources, page 326)
1 ounce Grand Marnier
1 ounce dry vermouth
4 dashes peach bitters, preferably Bitter Truth Peach Bitters
Dash of Angostura bitters
A fresh peach slice

Fill a cocktail shaker with ice. Add the Madeira, Grand Marnier, vermouth, peach bitters, and Angostura bitters and shake vigorously for 15 seconds. Strain into a chilled champagne saucer, garnish with the peach slice, and serve.

HOW TO SET UP A WHISKEY COCKTAIL STATION FOR A PARTY

I love inviting people to my home for cocktails. Entertaining is something that Southerners do well. And for the past several years, I have been collecting old bottles of bourbon, just like I used to collect baseball cards as a kid. Once you become a serious collector of a particular thing, you tend to study it. I would like to think by now that I know a thing or two about that old Southern corn whiskey that hibernates in charred oak barrels for years. I had floor-to-ceiling shelves built to display the hundreds of bottles that I have amassed over the years. I love seeing the look on people's faces when they see my "Great Wall of Bourbon" for the first time.

I'm happy to share my whiskey and passion with anyone brave enough to sit through the late-night whiskey "seminars" that I used to give. When I started the seminars, I would often line up ten to twelve different whiskeys after work and have people taste through them. This seemed like a great idea—what's a better way to teach people about bourbon? But I quickly realized that most people don't have the ability to drink whiskey neat and straight from the bottle like I do. So I stopped giving the seminars because my audience would get tanked and would barely be able to focus enough to listen to me ramble on.

Of course, I didn't want to stop sharing my love of fine bourbon with my friends. I feel like most people don't know what makes a bourbon great. So instead, I started making simple cocktails to give me the platform I need to wax poetic about mash bills and barrel proofs while keeping my audience a bit more sober. That worked a lot better, but it often meant I ended up stuck behind the bar pulling a bartender shift. So I came up with a way to set up a DIY bourbon bar instead. Here's how you can do it too.

Choose five bourbons that are different in flavor, age, and proof. Familiarize yourself with the individual bourbons. You'll want to know the mash bill (the grains used to make the bourbon), the proof (the alcohol content), and the age. The proof and the age will be written on the bottle and a quick Google search can tell you the specifics of each whiskey's mash bill and whether it has been softened with wheat, oats, rye, or barley. My go-to source is www.bourbonenthusiast.com. Fill out your station with glassware, water, ice, and various items like bitters, citrus (to squeeze or muddle or zest), and different sweeteners (see the list opposite).

WHAT YOU'LL NEED

A high-proof bourbon, something above 107 proof
A rye whiskey
A lower-proof bourbon (between 80 and 90 proof)
An older bourbon (at least 10 years old)
White corn whiskey

Agave nectar
Honey
Simple syrup
Raw sugar cubes
Cold spring water
A bucket of ice
Pok Pok drinking vinegars (see Resources, page 326)
Several kinds of bitters
Lemons
Oranges
Preserved cherries

Single-channel zester
Multi-channel zester
Vegetable peeler
Strainer
Large glass or crystal mixing pitcher, beaker, or Mason jar
Long cocktail spoons
Rocks glasses
Herbsaint, in an atomizer
At least three Junior Kimbrough albums

Have your guests make their own cocktails. Whiskey
cocktails are a very personal thing and this station will allow
for fun, experimentation, and learning.

Pick a whiskey and pour 2 ounces of it into a large
mixing glass. Taste it. Is it sweet? Mellow? Hot? What will
balance the whiskey to make it taste better?

HERE ARE SOME HINTS:

If it's hot, you'll want to add water and sugar.

If it's mellow, you'll want to accent it with acid or bitters
or both.

The higher-proof bourbons will benefit from some water
and a little sweetener. A higher-proof might also need
some Pok Pok vinegar added. The more subtle bourbons
may only need a dash of bitters and a twist of lemon.

White corn whiskey will be a blank slate and will take
brighter and lighter mixes (just as in a vodka drink). It'll
show how whiskey starts before it goes into the barrel.

Rye whiskey will taste of corn and rye. It's spicy, so mix it
with a sweetener and citrus to balance out the flavors.

If it's an older bourbon, you'll only want to add one thing
(not a sweetener!) or nothing at all. It'll be close to
perfect as is.

Every time you add a component, taste the cocktail. Did you
add too much of one thing? If so, you may need to balance
it with some more whiskey. Then stir, chill, strain it with ice,
pour your finished drink into a glass, and enjoy.

SOUTHERN SCREWDRIVER

A properly made screwdriver is pretty darn good. We upped the ante a bit, adding some spice by infusing the vodka with jalapeños and giving the drink a little herbal sweetness with a fresh basil simple syrup.

Chipped ice or ice cubes
2 ounces Jalapeño-Infused Vodka (recipe follows)
1 ounce vodka, preferably Smirnoff
2 tablespoons Basil-Infused Simple Syrup (recipe follows)
½ cup fresh orange juice
1 basil leaf
Jalapeño slices (optional)

Fill a cocktail shaker with ice, add both vodkas, the simple syrup, and orange juice, and shake vigorously. Strain into a chilled rocks glass over chipped or cubed ice. Smack the basil leaf between your palms to release the oils, garnish the drink with the leaf and jalapeño slices, if desired, and serve.

JALAPEÑO-INFUSED VODKA

MAKES 1 LITER

1 to 2 jalapeño peppers, depending on how spicy the peppers are and how hot you want the drink to be

One 1-liter bottle vodka, preferably Smirnoff

Slice the jalapeño pepper(s) lengthwise, leaving the seeds in, and drop them into the bottle of vodka. Seal the bottle and let the peppers infuse the vodka for 2 to 3 days. The longer the jalapeños sit in the vodka, the spicier it will become.

Strain the vodka to remove the jalapeños and seeds. Pour the vodka back into the bottle and seal the bottle. The vodka will keep indefinitely at room temperature.

BASIL-INFUSED SIMPLE SYRUP

MAKES 2 CUPS

1 cup sugar
1 cup hot water

4 large basil leaves

FOR THE SIMPLE SYRUP: Combine the sugar and water in a small saucepan over high heat and stir until the sugar dissolves. Cool to room temperature, pour into a clean quart jar, and refrigerate until chilled.

Put ¼ cup of the chilled syrup and the basil in a blender and blend on high until smooth, about 3 minutes. Stir the puree back into the remaining simple syrup. Tightly sealed, the syrup will keep for up to 3 days in the refrigerator.

THE CHARLESTON LIGHT DRAGOON PUNCH 1792

MAKES 20 SERVINGS

The plight of the Light Dragoons is a tragic Charleston story. Back when men assembled militias to defend a city, the Light Dragoons were some of the wealthiest of the bunch. Formed in 1792, they charged dues and required members to supply their own mounts. Most of them were slaveholders and growers of rice and cotton. For the majority of the company's existence, the dragoons mounted their horses for ceremonial purposes only, to lead parades and such about town. But in 1864, the company became mired in several Civil War battles. Within a few short weeks, almost all the men were dead. Of the sixty or so who made the trip to Virginia to fight, only a handful returned.

A friend found this recipe at the South Carolina Historical Society and updated it. The original recipe was from the militia.

2 quarts water

7 bags black tea, preferably American Classic (see Resources, page 326)

2 cups raw sugar

1½ cups fresh lemon juice

12.7 ounces brandy (California is fine)

12.7 ounces rum, preferably Cockspur Barbados

6.4 ounces peach brandy

20 large ice cubes

2¾ cups soda water

20 thin slivers of lemon peel (from about 3 lemons)

Bring the water to a boil in a medium stainless steel saucepan over high heat. Add the tea, remove the pan from the heat, and steep the tea for 20 minutes.

Strain the tea through a tea strainer or a fine-mesh strainer into a 1-gallon container. Add the sugar and stir until it is completely dissolved. Let the mixture cool to room temperature, about 20 minutes.

Add the lemon juice, brandy, rum, and peach brandy to the tea mixture, cover, and refrigerate until cold. (Tightly covered, the punch base will keep for up to 3 days in the refrigerator.)

To serve, ladle 3 ounces of the punch base into each punch cup. Add an ice cube, top off with 1½ ounces of soda, and garnish with a sliver of lemon peel.

THE PECAN WHISKEY DAISY

MAKES 1 DRINK

The Whiskey Daisy first appeared in 1888, in Harry Johnson's famous *Bartenders' Manual*. The recipe called for chartreuse, lime, and lemon flavors, but nowadays there are a bunch of different variations of this cocktail. At some point orgeat, a syrup made with orange water, almonds, and grain alcohol, found its way into the drink. Our version gets a Southern accent by using pecans instead of almonds. I prefer rye whiskey for this cocktail, but you can use your favorite spirit.

Chipped ice
3½ ounces rye whiskey
½ ounce Pecan Orgeat (recipe follows)
1 tablespoon fresh lemon juice
2 tablespoons Raspberry Syrup (recipe follows)

Fill a cocktail shaker with ice. Add the whiskey, orgeat, lemon juice, and raspberry syrup and shake hard for 10 seconds. Strain into a champagne saucer over chipped ice and serve.

PECAN ORGEAT

MAKES 3 CUPS

3 cups water
1¾ cups raw pecans
¾ cup sugar

½ ounce 151-proof grain alcohol, such as Everclear
¾ teaspoon orange blossom water (see Resources, page 326)

Bring the water to a boil in a medium saucepan. Add the pecans, remove the pan from the heat, and let the pecans steep for 10 minutes.

Drain the pecans, reserving half of the water. Place the pecans and reserved water in a food processor and pulse several times, until the nuts are coarsely chopped. Transfer to a container and let cool to room temperature, about 10 minutes.

Set a fine-mesh sieve over a saucepan and line it with a double layer of cheesecloth. Drain the pecans and liquid in the cheesecloth, letting the liquid drip into the saucepan. Twist the cheesecloth to get the last bit of liquid; discard the cheesecloth and pecans. Add the sugar to the pecan liquid and bring to a simmer over medium-high heat, stirring to dissolve the sugar. Remove the pan from the heat and allow the mixture to cool to room temperature.

Pour the mixture into a nonreactive container and stir in the grain alcohol and orange blossom water. Cover and refrigerate. Tightly covered, the orgeat will keep for up to 1 week in the refrigerator.

RASPBERRY SYRUP

MAKES 1½ CUPS

1¾ cups fresh raspberries
½ cup sugar

¾ cup water
1½ teaspoons fresh lemon juice

Puree the raspberries in a blender.

Strain the raspberries through a fine-mesh sieve into a small nonreactive saucepan; you simply want to remove the seeds. Add the sugar, water, and lemon juice, stir, and bring the mixture to a boil over medium-high heat. Boil, stirring once or twice, for 2 minutes.

Strain the syrup through a fine-mesh sieve into a heat-proof container and cool to room temperature, then cover and refrigerate. Tightly covered, the syrup will keep for up to 1 week in the refrigerator.

FIRE IN THE ORCHARD

We only make this cocktail when we can get good apples in season, which means an annual disappointment for its fans when they disappear from the market in the warmer months. The combination of fruit-laden liquor and the sharp bite of capsicum peppers makes people sit up straight and ask for another, and perhaps another. The smoked apple juice for the cocktail is delicious on its own too.

1 Granny Smith apple

1 raw sugar cube

Dash of bitters, preferably Boker's Baked Apple Bitters (see Resources, page 326)

Slice of jalapeño pepper

Chipped ice or ice cubes

1½ ounces bourbon

1 ounce Laird's Applejack Brandy

Splash of Cointreau

A few drops of the brine from pickled jalapeños

Bourbon-soaked cinnamon stick

EQUIPMENT

Juice extractor

Smoker

Hickory wood chips, soaked

Run the apple through a juicer; you only need 3 tablespoons of juice, but smoke it all and drink the rest (or make another cocktail).

Smoke the apple juice at 75°F to 125°F for 6 hours, until the juice has a strong smoke flavor. Let cool. *(The smoked juice can be prepared up to 3 days ahead, cooled to room temperature, covered, and refrigerated for up to 3 days.)*

Put the sugar cube, bitters, and slice of jalapeño pepper in a cocktail glass and muddle them with a muddler or spoon.

Fill a cocktail shaker with ice, add the bourbon, brandy, Cointreau, jalapeño brine, and 3 tablespoons of the smoked apple juice, and shake vigorously for 10 seconds. Strain into the glass over the ice. Garnish with the bourbon-soaked cinnamon stick and serve.

HERBAL TEA COCKTAIL

We have experimented with infused vodkas for years at the bars at McCrady's and Husk. We've made vodka that tastes like sweet Southern tea, vodka that tastes like beets, and vodka that tastes like Bing cherries. That's the wonderful thing about vodka; it functions as a canvas on which a barman can add many layers of flavor. For this cocktail, we married a rather floral expression of various herbal blossoms into a concoction that could probably cure the common cold, albeit with a decidedly sharp kick!

Chipped ice

3 ounces Herb-Infused Vodka (recipe follows)

½ ounce Raspberry Syrup (page 326)

Splash of Cointreau

Fill a cocktail shaker with ice, add the vodka, raspberry syrup, and Cointreau, and shake vigorously for 10 seconds. Strain into a chilled rocks glass and serve.

HERB-INFUSED VODKA

¾ cup dried organic chamomile blossoms	¾ cup dried organic hibiscus blossoms
¾ cup dried organic lavender flowers	¾ cup chopped dried orange peel
	One 1-liter bottle vodka, preferably Smirnoff

Wrap the chamomile, lavender, hibiscus, and orange peel in a square of cheesecloth, tie the sachet, and add it to the bottle of vodka. Seal the bottle and let the herbs infuse the vodka for 3 weeks.

Remove the sachet and discard. Seal the bottle. The infused vodka will keep for 2 weeks at room temperature.

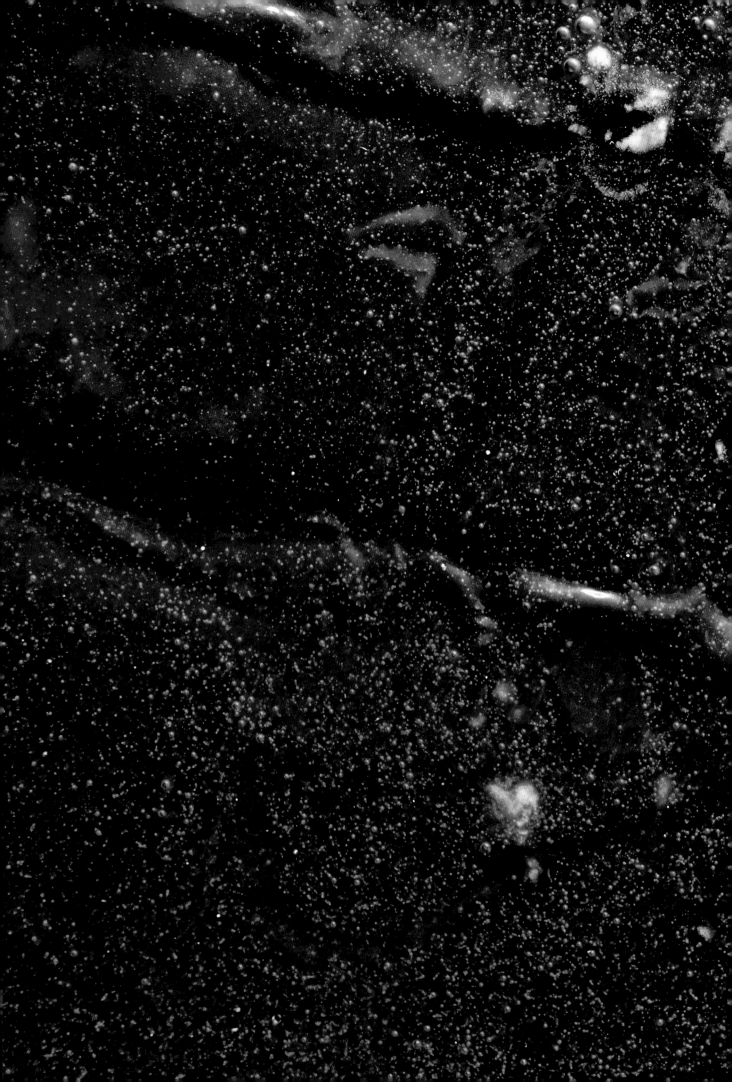

THE SWEET KITCHEN

SOUTHERN DESSERTS, A TRADITION

Southerners certainly have a sweet tooth. You'll already know this if you've ever tasted a glass of our iced tea. And Southerners view desserts as a way to both pass down the old traditional ways and share stories of family members and times past.

These traditions evolved within communities and among families. We all have favorite desserts, ones that we recall from our childhood, from family reunions and holidays.

I'm no pastry chef and will never claim to be one. I decided a long time ago that I would leave the baking to the experts, But I can promise you that I never skip dessert. In fact, I get downright giddy at the thought of ice cream, cake, and pie. Because my fondest memories of desserts are enjoying them at home, I decided not to include restaurant desserts in this book. You won't find our complicated McCrady's desserts here or recipes made with fancy kitchen equipment. What you will find are the desserts that I make at home to serve to my family and friends and recipes for the simple desserts you'll find on the menu at Husk.

Some of my favorite heirloom ingredients are found in the sweet kitchen. The dessert canon is no different from anything else in the South in terms of using what we grow. When I was a kid, the desserts I ate were made with sorghum and black walnuts and blackberries, because that's what grew, what was harvested, and what was available at arm's length. I carry on this tradition. Benne shows up in my desserts in the form of the batter for Antebellum Benne–Blackberry Tart (page 282). I use Carolina Gold rice to make Calas (page 298) and rice pudding (see page 297). Sorghum is used in my grandmother's stack cake (see page 284) and in bacon caramels (see page 303). And you will find my favorite black walnuts in the Hillbilly Fudge (page 305) and Black Walnut Pound Cake (page 278).

These recipes are special to me because they're the ones I've come to expect from certain family members at holiday gatherings, where everyone has their own dessert that they're known for: my aunt's pumpkin roll, my sister's chocolate éclair cake, my gran's stack cake. In the end, these traditions tie us together as a family and as proud Southerners as well, and we make sure to make the ending of a meal just as important as the beginning.

BLACK WALNUT POUND CAKE
WITH CHOCOLATE GRAVY

SERVES 16

I created this cake in honor of my grandmother, who called the chocolate sauce "gravy." She's the reason I am a chef, and she's completely responsible for my passion for cooking. The cake resonates with the deep flavor of the black walnuts. When you make this cake, be sure to use the American black walnut, not the larger less assertive English walnut, the type commonly found in grocery stores.

CAKE

3 cups all-purpose flour

1 teaspoon baking powder

½ pound (2 sticks) unsalted butter, at room temperature

¾ cup Rendered Fresh Lard (page 316), at room temperature

1 tablespoon black walnut oil (see Resources, page 326)

2 cups sugar

5 large eggs

½ cup whole milk

½ cup whole-milk buttermilk

1 teaspoon pure vanilla extract

1½ cups chopped black walnuts (see Resources, page 326)

GRAVY

½ cup sugar

¼ cup unsweetened cocoa powder

3 tablespoons all-purpose flour

2¼ cups whole milk

2 tablespoons unsalted butter, diced

2 teaspoons Bourbon Barrel Foods Aged Vanilla Extract (see Resources, page 326)

EQUIPMENT

10-inch tube pan

FOR THE CAKE: Position a rack in the middle of the oven, with no rack above it, and preheat the oven to 325°F. Grease and flour a 10-inch tube pan.

Sift the flour and baking powder into a small bowl and set aside.

In the bowl of a stand mixer fitted with the paddle attachment, or in a large mixing bowl, using a hand mixer, cream the butter, lard, oil, and sugar together until light and fluffy, about 3 minutes. Add the eggs one at a time, beating well after each one is added. Combine the milk, buttermilk, and vanilla in a small bowl and add to the creamed mixture, beating on low speed just until combined. Add the flour mixture and beat just to combine. Fold in the nuts.

Pour the batter into the prepared pan. Tap the pan on the counter to level the top of the batter. Bake the cake for 1 hour and 25 minutes, or until a cake tester inserted in the center comes out clean. Cool for 10 minutes on a baking rack before removing the cake from the pan. Let cool completely on the rack.

FOR THE GRAVY: Put the sugar, cocoa powder, and flour in a medium saucepan and whisk to combine and break up any lumps. Whisk in the milk. Put the saucepan over medium heat and cook the mixture, stirring occasionally with a silicone spatula, until it thickens, 7 to 10 minutes. When the mixture is as thick as pancake batter, remove it from the heat. Stir in the butter one piece at a time until incorporated. Add the vanilla. *(The gravy can be cooled to room temperature, covered, and refrigerated for up to 3 days; warm before serving.)*

Slice the cake and serve with some of the warm gravy over each slice.

In a cake keeper or cake box, the cake will keep for up to 1 week at room temperature. It can also be frozen for up to 3 months. Wrap it securely in plastic wrap and then in aluminum foil.

BLACK WALNUTS

When I was growing up, we always had a black walnut tree nearby and every fall the ground would be covered with those green orbs that stained your hands for days when you touched them. As a kid, I thought it was funny to throw them at my friends, ruining T-shirts and then finding ourselves in trouble with our mothers. Now that I appreciate black walnuts for their intense bitter flavor and tannic bite, I can't imagine wasting them. I fondly remember sitting with my grandmother under her walnut tree with a hammer and a couple of cinder blocks at our sides. We'd stain our fingers a dark purple from busting the hard shells of the black walnuts (which are also used to make a natural dye).

Black walnuts are primarily used in sweets, where they can add a mysterious tang to a batch of rich fudge (see page 305) or be the main star in a deliciously moist pound cake (see page 278). The American black walnut has a bitterness and deep flavor that the European walnut lacks. You can swap in the grocery store variety in the recipes in this chapter, but I think you owe it to yourself to order a bag of the real thing.

The black walnut grows throughout the central South, but its numbers have been drastically affected by the value of its wood. Prize specimens can go for several thousand dollars—enough money that there are even walnut poachers who chop down huge trees and cart them off in the dead of night. The wood is strong, so it's a favorite for makers of furniture and gunstocks. So the black walnut is scarcer these days. But if you have access to some woods, the black walnut tree can be found in areas with deep, rich soils; you'll want to search the valley, not the mountaintop. And if you're not up for tromping through the woods, there is a company called Hammons (see Resources, page 326) that gathers walnuts from hundreds of individual growers and sells them online. You can get them cracked or whole. I suggest getting some whole ones to fully understand why people prize the meats. Just watch your fingers when you crack them open!

MY SISTER'S CHOCOLATE ÉCLAIR CAKE

MAKES EIGHTEEN
3-BY-2¼-INCH PIECES

My sister made chocolate éclair cake for our Thanksgiving dinner the first year she moved out of the house. I loved it so much that she makes it for me every year for my birthday. It's become such a family tradition that now her daughter is making it too. The version I grew up on included store-bought graham crackers, Jell-O pudding mix, and Cool Whip, which is probably why I loved it so much. My version uses Anson Mills Graham Flour and home-made vanilla pudding and whipped cream. You can make it either way.

CRUST

1½ ounces whole milk

¼ cup sorghum

½ teaspoon pure vanilla extract

1½ cups Anson Mills Antebellum-Style Rustic Coarse Graham Flour (see Resources, page 326)

½ cup Anson Mills Pizza Maker's Flour (see Resources, page 326)

½ cup packed dark brown sugar

¾ teaspoon baking powder

½ teaspoon baking soda

½ teaspoon kosher salt

⅛ teaspoon ground cinnamon

6 tablespoons unsalted butter, cubed and chilled

VANILLA PUDDING

4 cups whole milk

½ cup heavy cream

1 vanilla bean

1 tablespoon plus 1 teaspoon pure vanilla extract

1 cup sugar

½ cup cornstarch

½ teaspoon kosher salt

6 large egg yolks

6 tablespoons unsalted butter, at room temperature

VANILLA CREAM

1 sheet silver-strength gelatin (see Resources, page 326)

2 tablespoons water

3 cups heavy cream

1 cup confectioners' sugar

1 teaspoon pure vanilla extract

24 graham crackers

GLAZE

1 cup sugar

¼ cup whole milk

⅓ cup unsweetened cocoa powder

2 tablespoons unsalted butter

1 teaspoon pure vanilla extract

FOR THE CRUST: Combine the milk, sorghum, and vanilla in a medium bowl and whisk to combine. Set aside.

Combine the flours, brown sugar, baking powder, baking soda, salt, and cinnamon in a food processor and pulse several times to combine. Add the cold butter and pulse 8 to 10 times until the mixture has the texture of coarse corn-meal. With the food processor running, slowly stream in the liquids. Continue processing until the mixture forms a ball, 1 to 2 minutes.

Turn the dough out onto a lightly floured surface. Pat into a flat, round shape. Wrap in plastic wrap and refrigerate for 30 minutes.

Preheat the oven to 350°F. Line a 9-by-13-inch baking pan with parchment paper or a Silpat.

Roll the dough out between two sheets of parchment paper to a ¼-inch thickness. Cut into 2-inch squares and trans-fer them to the baking pan, leaving 1½ inches between them.

Bake for 10 to 12 minutes, until evenly brown. Allow to completely cool in the pan set on a baking rack.

FOR THE VANILLA PUDDING: Combine 2 cups of the milk and the cream in a medium heavy-bottomed saucepan and heat over medium heat. Split the vanilla bean lengthwise in half and scrape out the seeds, using the back of a paring knife. Add them and the vanilla to the mixture. Put the sugar, cornstarch, and salt in a medium bowl and whisk to combine. Add the remaining 2 cups milk and whisk in the egg yolks.

When the milk mixture starts to steam—*not* boil—use a ladle and whisk to slowly temper the cornstarch slurry into the pan by pouring cups of the milk mixture into the slurry and whisking vigorously for 30 seconds. Pour all the contents of the bowl into the saucepan. Cook over medium heat, stir-ring constantly, until the mixture thickens significantly, 12 to 15 minutes.

Remove from the heat and strain through a fine-mesh sieve or chinois, pressing the pudding with the back of a spatula. Whisk the softened butter into the pudding until com-pletely incorporated. Transfer to a storage container and press plastic wrap directly on the pudding to avoid a skin forming on the top. Refrigerate for at least 2 hours.

MEANWHILE, FOR THE VANILLA CREAM: Place the gelatin sheet and 1 cup cold water in a small bowl and let the gelatin soften, 5 to 10 minutes. Meanwhile, warm the 2 tablespoons water in a small saucepan. When the gelatin sheet is soft, lift it from the water and gently wring it out. Add it to the warm water and heat over medium-low heat, stirring, until the gela-tin has dissolved, about 1 minute. Remove the saucepan from the heat.

In the bowl of a stand mixer fitted with the whisk attachment, or in a large mixing bowl, using a hand mixer, whisk the cream with the confectioners' sugar and vanilla until it reaches soft peaks, 3 to 4 minutes. Stream in the lukewarm gelatin mixture and continue whisking until the cream reaches a medium-stiff peak.

TO BEGIN TO ASSEMBLE: Line a 9-by-13-inch baking pan with parchment paper. Fold the vanilla cream and the chilled pudding together.

Place a layer of graham crackers evenly on the bottom of the prepared pan. Spread half of the vanilla-cream-pudding mixture over the crackers, and top with another layer of grahams. Spread the remaining mixture on top and finish with another layer of crackers. Refrigerate to chill, at least 1 hour.

MEANWHILE, FOR THE GLAZE: Combine the sugar, milk, and cocoa powder in a small saucepan and heat over medium heat, stirring constantly, for 1 minute. Remove from the heat, add the butter and vanilla, and whisk to combine.

TO ASSEMBLE: Pour the warm glaze over the entire cake (still in its pan) and chill again to set, 1 to 2 hours.

The cake can be stored, covered, in the refrigerator for up to 3 days.

ANTEBELLUM BENNE–BLACKBERRY TART
WITH BROWN BUTTER ICE CREAM

Bitter but sweet. That's the first thing that comes to mind when I think of this dessert. The pleasant bitterness of the antebellum benne riffs against the natural sweetness of super-ripe blackberries. It's a perfect flavor combination. Best of all, this recipe is easy to make, and you can customize it to use whatever fruit is ripe, in season, and ready to eat. I make the tart with everything from wild plums to persimmons. Serve it with the brown butter ice cream, which is awesome all by itself too.

Note that the ice cream base needs to chill for a total of 48 hours before freezing.

BROWN BUTTER CRUMBS

½ pound (2 sticks) unsalted butter, diced

1 cup dried nonfat milk powder

BLACKBERRY TART

1 pound (2 cups) blackberries, washed and dried

¾ pound (3 sticks) unsalted butter, diced

2 vanilla beans, split lengthwise

1 teaspoon kosher salt

12 large egg yolks

1½ cups sugar

2 tablespoons whole-milk buttermilk

1 teaspoon baking powder

½ cup all-purpose flour

½ cup Anson Mills New Crop Heirloom Bennecake Flour (see Resources, page 326)

2 tablespoons Anson Mills Antebellum Benne Seeds (see Resources, page 326)

Brown Butter Ice Cream (recipe follows)

FOR THE CRUMBS: Melt the butter in a medium heavy-bottomed saucepan over medium heat. Add the milk powder and cook, stirring constantly, until the butter is golden brown and starts to smell slightly nutty, about 5 minutes. (The milk powder will want to stick to the pan and can scorch, so it is critical to keep the mixture moving.) Pour onto a rimmed baking sheet and let cool.

Once it is cool, put the butter mixture in a blender or food processor and pulse until crumbly. *(The crumbs can be stored for up to 2 days in an airtight container.)*

FOR THE TART: Position a rack in the middle of the oven and preheat the oven to 350°F. Line a 9-inch cast-iron skillet (or a pie plate) with a round of parchment paper.

Arrange the blackberries in the skillet in a single layer; set aside.

Put the butter, vanilla beans, and salt in a medium heavy skillet and cook over medium heat, stirring, until the butter is golden brown and starts to smell slightly nutty, about 5 minutes. Strain the butter through a fine-mesh strainer into a small bowl; discard the solids and vanilla beans. Allow the butter to cool to room temperature.

Whisk the yolks and sugar in a large mixing bowl until thick and light yellow. Gradually whisk in the cooled brown butter. Whisk in the buttermilk. Combine the baking powder and the two flours in a large bowl, then gradually whisk into the yolk mixture to form a thick batter.

Pour the batter over the blackberries. Smooth the top with an offset spatula. Bake the tart for 30 minutes. Rotate it back to front and bake it for 20 to 30 minutes more, until a toothpick inserted into the center comes out clean. Cool it in the pan for about 20 minutes.

Slide the tart out of the pan and cut it into 10 servings. Sprinkle each slice with some brown butter crumbs and benne seeds and serve with a scoop of ice cream.

Tightly covered, the tart will keep for up to 2 days in the refrigerator. Bring it back to room temperature before serving.

BROWN BUTTER ICE CREAM

MAKES 3 QUARTS

1 pound unsalted butter, diced

3 cups whole milk

3 cups heavy cream

15 large egg yolks

1½ cups sugar

¾ teaspoon kosher salt

EQUIPMENT

Ice cream maker

Put the butter in a heavy skillet and cook over medium heat, stirring, until it is golden brown and starts to smell slightly nutty, about 5 minutes. Strain the butter through a fine-mesh strainer into a medium bowl; discard the solids. Pour in the milk and cream, cover the bowl, and refrigerate overnight.

The next day, there will be a solid layer of butter on top of the milk mixture. Make a hole in the layer of butter and pour the liquid under it into a large saucepan. Discard the butter or enjoy it as a snack. Bring the liquid to 180°F over medium heat, stirring often so that it doesn't scorch, about 8 minutes.

Meanwhile, put the egg yolks in a large bowl and whisk in the sugar and salt. Remove the saucepan from the heat. Slowly stream 1 cup of the hot liquid into the egg and sugar mixture, whisking constantly. Add another cup, whisking. Repeat until all the liquid is added. Strain the custard through a fine-mesh sieve into another large bowl.

Make an ice bath in a bowl with equal parts ice and water. Put the bowl of custard in the ice bath to chill and stir occasionally, being careful not to let any water get into the custard, until the custard is completely cold. Cover it and refrigerate overnight.

The next day, pour the custard into an ice cream maker and freeze, following the manufacturer's instructions. Transfer the ice cream to a freezer container and freeze until you are ready to serve it.

Tightly covered, the ice cream will keep for up to 1 month in the freezer.

AUDREY MORGAN'S APPLE-SORGHUM STACK CAKE

SERVES 10 TO 12

One sniff of an apple-sorghum stack cake gets me so excited I can barely stand it. As a kid, I never really liked this beast of a cake: it wasn't colorful, and it wasn't coated in super-sweet icing. I just couldn't understand why the adults were so crazy about it. When I got older and started to appreciate food a little more, I also started to appreciate the work that goes into making a stack cake.

I adapted this cake from my grandmother's handwritten recipe. When I first made it, I wanted everyone to taste how brilliant this Southern masterpiece was, and I was surprised at how many people had never even seen a stack cake before. Then I gave a piece to a cook named Maya. She took a bite and I watched her tear up: "Oh, my God, that tastes just like the one my granny used to make."

If you don't have springform pans, use two 9-inch round cake pans and bake the layers two at a time. This recipe makes twelve thin, delicate layers. If one of two break and become unusable, call them the baker's treat. Note that the glazed cake must sit overnight before serving.

APPLE BUTTER

27 cups chopped unpeeled apples (about 18 apples), preferably Pippin or Granny Smith

1⅔ cups apple cider

5 cups sugar

1½ tablespoons ground ginger

2¼ teaspoons ground cinnamon

2¼ teaspoons grated nutmeg

1 teaspoon ground cloves

CAKE

8 tablespoons (1 stick) unsalted butter, at room temperature for greasing the pans

9 cups self-rising flour, preferably White Lily

2 cups sugar

4 teaspoons ground ginger

2 teaspoons ground allspice

2 teaspoons ground cinnamon

6 large eggs

1½ cups canola oil (my grandmother used shortening; lard [see page 316] would be tasty as well)

2 cups whole-milk buttermilk

2 cups sorghum (see Resources, page 326)

GLAZE

2 cups packed light brown sugar

½ cup rye whiskey or bourbon

One 14-ounce can sweetened condensed milk

8 tablespoons (1 stick) unsalted butter, diced

½ cup whole milk

1 teaspoon pure vanilla extract

EQUIPMENT

One or more 10-inch springform pans or two 9-inch cake pans

FOR THE APPLE BUTTER: Combine the apples and cider in a large heavy-bottomed nonreactive pot, cover, and cook over low heat, stirring, until very soft, about 4 hours.

Pass the apples through a food mill and return to the pot. Add the sugar and spices, cover, and cook the apple butter over low heat, stirring frequently, until it is very thick, 1 to 2 hours. Remove from the heat. You need 6 cups apple butter for the cake. Transfer to a container and cool to room temperature, then cover and refrigerate. *(Tightly covered, the apple butter will keep for several weeks in the refrigerator, or it can be frozen for up to 3 months.)*

FOR THE CAKE: Position a rack in the middle of the oven and preheat the oven to 350°F. You will need to make 6 cake layers. Butter the bottoms and sides of the pans, line each with a parchment circle, and butter the parchment.

Sift the flour, sugar, ginger, allspice, and cinnamon into a large mixing bowl and combine well.

In a medium bowl, lightly beat the eggs, then gently whisk in the oil, buttermilk, and sorghum; you do not want to make a lot of froth. Slowly stir the wet mixture into the dry mixture.

Pour 2 cups of batter into each prepared pan and spread it out using the back of a spoon. Bake for 16 minutes, or until a toothpick inserted in the center comes out dry. Allow the layers to cool in the pans for 10 minutes and then turn out onto baking racks. Once they are cool, peel off the parchment paper. Use the pan(s) for the next layer(s) and repeat.

Remove the apple butter from the refrigerator and let come to room temperature. Using a long serrated knife, carefully split each layer in half as follows. Make 4 evenly spaced horizontal marks around the layer, halfway up. Slice a few inches in toward the center from one mark, turn the layer, and slice in at the next mark, and continue. The cuts should meet at the center so that you get 2 layers. Once you have sliced through to the center, lift off the top layer and place it on a work surface. Repeat with the remaining 5 layers.

Place one layer cut side up on a cake plate. Spread a heaping ½ cup of apple butter evenly over it. Place a second layer on top of the apple butter, cut side down, and spread another ½ cup of apple butter over it. Continue the process with the remaining layers of cake, leaving the last layer plain.

FOR THE GLAZE: Combine the brown sugar, whiskey, and condensed milk in a medium saucepan and cook over medium heat, stirring constantly, until the sugar dissolves. Remove from the heat and stir in the butter until it is melted. Stir in the milk and vanilla. Cool on the counter.

Pour the glaze over the cake, allowing it to run down the sides. Put the cake in a cake box and let set overnight at room temperature before cutting. Covered, the cake will keep for up to 3 days at room temperature or 5 days in the refrigerator.

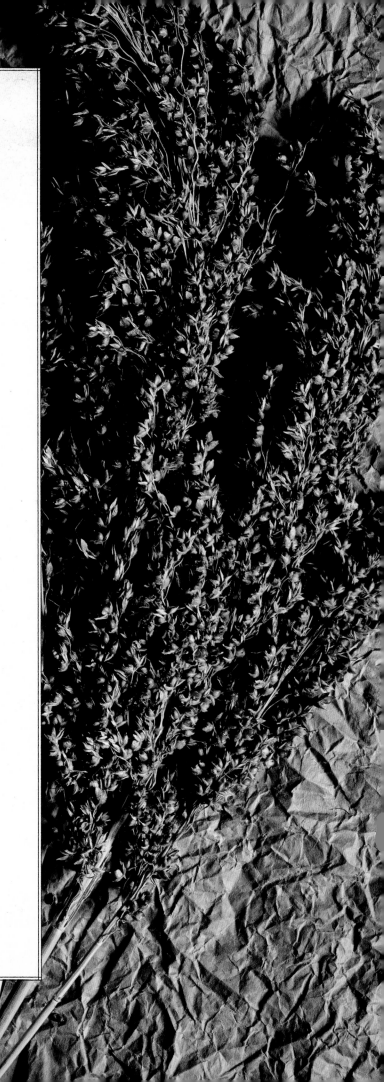

SORGHUM

Sorghum is a unique crop and a lot of people think that they have had sorghum when they've really been using molasses. Molasses is a by-product of the sugar-refining process. Syrup made from the sorghum plant, which is a variety of millet originally from Africa, has a completely different flavor, deep and woodsy.

In the temperate regions of the Old South, before modern industry and shipping, sorghum was often the only affordable source of sweetness for many folks. The sticky syrup was poured on biscuits or just a piece of toast with homemade butter. Sorghum is still used in desserts today. I pour it over my Calas (see page 298) and use it to flavor caramels (see page 303), and you can make cookies, pies, and all sorts of other sweets with it.

Some of the first sorghum plants grown in the United States were planted by Governor James Henry Hammond in what is now Barnwell County, South Carolina. His experiments in the 1850s with what was then known as "imphee" produced good results, and word soon spread that this new plant could make sweet syrup. The same stalk that could be squeezed for juice also produced millet—a grain that can feed livestock as adequately as corn. So sorghum represented a double yield for farmers: juice for the syrup kettle and grain for the hogs. Farmers from Georgia to Ohio would plant sorghum and then come together during the fall harvest to boil down the juice.

The flavors and smells of sorghum remind me of home. Sorghum brought my local community together; people gathered in groups to spread the labor of cooking down the hundreds of gallons of juice squeezed from the green stalks. We pressed the gathered sorghum stalks through a large mill together. A circling mule powered it. Then a fire was built under a kettle and the juice was cooked down in batches into syrup. I would remove the scum that floated to the top of the kettle with a tool made from a perforated tin dustpan stuck onto the end of a broom handle. At the end of a long day, the efforts were split up between the neighbors and that was our sorghum supply for the year.

All across the South, there are families who keep this practice alive, as much for the camaraderie as for the jars of syrup that everyone goes home with. If you're a farmer or have a few acres to spare, I suggest researching the syrup-making process. Perhaps a few people will have enough interest to buy and restore an old mill, brick up a hundred-gallon syrup kettle, plant a stand of sweet millet, and help to preserve this dying art.

SCOTT WITHEROW

Scott Witherow is a true Southern chocolatier. When I was first working in Nashville, he liked the whole molecular gastronomy bit and would come hang out at Capitol Grille. We would always end up at the same bar together getting drunk and talking food. He had come back from England with a culinary degree and was working his way through kitchens in town—and the guy had one hell of a sweet tooth. Today Scott is the most accomplished chocolatier in the South. He and his chocolates are one of a kind. By blending Southern sensibilities and ingredients with traditional European techniques (and machines), he is turning out some of the most creative sweets in the country.

Scott started his chocolate business in his own home, but bean-to-bar chocolates aren't easy to make in a home kitchen. He had to be inventive, using, for example, a blow dryer and an old chinois to winnow off the cocoa bean chaff. But pretty soon he was showing up all over Nashville, getting chefs to taste his experiments with various chocolate techniques and to critique his progress.

By his own admission, the first batches needed a lot of work. But Scott stuck with it. He started buying all sorts of weird old chocolate machines from dealers in Europe, fixing them up himself, and producing bars of fine chocolate infused with uniquely Southern flavors.

Once the chocolate was good enough, Scott opened Olive and Sinclair, a boutique shop on the east side of Nashville. He's the only guy I know of who uses brown sugar to sweeten his chocolate; you can taste the molasses in every bite. Buttermilk flavors his signature white chocolate, and he uses cocoa nibs smoked by Allan Benton in one of his best-selling chocolate bars. His chocolates are strong and brash and insanely delicious.

Guys like Scott succeed because they have passion. He is obsessed with his art and intently focused on quality. Scott is my counterpart in the sweet kitchen. The man is dedicated to Southern artisanal ingredients. We share a devotion to blending Southern culture into new mediums and establishing our culinary heritage as one of the world's great cuisines.

CHOCOLATE ALABAMA STACK CAKE

SERVES 10 TO 12

A couple of years ago, I became infatuated with stack cakes. They aren't traditional layer cakes—they look more like a stack of pancakes piled up with a sweet filling between the layers. Unless you have a cousin in the hills of East Tennessee or the fields of lower Alabama, stack cakes can be hard to come by these days. The story goes that people in the South were so poor that wedding cakes were built out of layers each guest brought to the ceremony—the more layers, the more popular the couple, I guess! But it's more likely that stack cakes required less equipment to produce and could be more easily packed for travel or picnicking, which has always been a very Southern pursuit. I've also heard that you can measure the skill of the baker by the tenderness of the crumb and how many layers the cake has, but I just find them fun to make.

CAKE

½ pound (2 sticks) unsalted butter, diced, at room temperature, plus more for the pans

3 cups all-purpose flour

¼ teaspoon kosher salt

1 heaping teaspoon baking powder

12 ounces 67% bittersweet chocolate, preferably Olive and Sinclair (see Resources, page 326), chopped

2 cups sugar

5 large eggs

1 cup evaporated milk

½ cup water

2 teaspoons pure vanilla extract

ICING

2 cups sugar

1 cup evaporated milk

5 ounces unsweetened chocolate, chopped

8 tablespoons (1 stick) unsalted butter, cut into chunks

1 teaspoon pure vanilla extract

EQUIPMENT

At least two 9-inch round cake pans, preferably more

FOR THE CAKE: Position a rack in the middle of the oven and preheat the oven to 350°F. Lightly butter as many 9-inch round cake pans as you have—you want to create a total of 10 layers, baking 2 or 3 layers at a time. If you only have two or three pans, cool them between batches and butter them again as needed.

Sift the flour, salt, and baking powder into a small bowl and stir to combine well. Set aside.

Put the chocolate in the top of a double boiler. Fill the bottom of the double boiler with water, insert the top, and set the double boiler over low heat; the water should never be hotter than a simmer. Stir the chocolate with a silicone spatula until it melts, scraping down the sides as necessary. Remove from the heat.

In the bowl of a stand mixer fitted with the paddle attachment or in a large mixing bowl, using a hand mixer, beat the butter and sugar on medium speed until light and creamy, about 5 minutes. Add the eggs one at a time, beating until smooth after each addition. Reduce the speed to low and add the sifted dry ingredients 1 cup at a time, beating until incorporated. Add the evaporated milk and melted chocolate and beat just to combine. Add the water and vanilla, beating until well combined.

Place a scant ⅔ cup batter in each prepared cake pan and use the back of a spoon or an offset spatula to spread the batter evenly. Bake 2 or 3 layers at a time for 8 to 9 minutes—a layer is done when you hold it near your ear and do not hear it sizzle. Allow the layers to cool in the pans for 10 minutes. Turn them out onto baking racks, using a metal spatula to ease them out of the pans, and let cool completely.

MEANWHILE, FOR THE ICING: Combine the sugar and evaporated milk in a medium saucepan and set over medium-low heat. Add the chocolate and butter and heat, stirring, until they have melted. Increase the heat to medium and cook, stirring occasionally, for 10 minutes, until the mixture has emulsified and is smooth. Remove from the heat and add the vanilla, stirring to combine. The icing will be thin, but it will thicken as it cools. Let cool completely.

TO ASSEMBLE: Place one layer on a cake plate and, using an icing spatula, spread 2 or 3 spoonfuls of icing on top. Repeat with the remaining layers (don't worry if a layer tears; no one will see it when the cake is finished). Then cover the top and sides of the cake with the remaining icing. Scrape up any icing that runs onto the plate and spread it back on the cake. If the icing hardens before the cake is iced, place the icing back over low heat.

Let the cake stand for 1 hour before serving it.

In a cake keeper or cake box, the cake will keep for up to 3 days at room temperature or 5 days in the refrigerator.

CHOCOLATE CHESS PIE

I was looking for the perfect way to show off Scott Witherow's chocolate when I developed the recipe for this chocolate chess pie. Whenever it graces the menu at Husk, it sells out almost instantly. Chess pie has a pretty diverse history in the South. It seems like everyone has a story of its origin and their own way of making it. My version is pretty darn sweet. If sweet is not your thing, add a teaspoon of distilled white vinegar to the filling for a little tang.

CHOCOLATE CRUST

1½ cups all-purpose flour, plus more for the work surface

¼ cup unsweetened cocoa powder

1 tablespoon sugar

1 teaspoon kosher salt

2½ sticks (10 ounces) unsalted butter, cubed and chilled

¼ cup plus 3 tablespoons ice water

FILLING

4 tablespoons unsalted butter

1½ ounces 67% bittersweet chocolate, preferably Olive and Sinclair (see Resources, page 326)

1½ cups sugar

1 tablespoon unsweetened cocoa powder

½ teaspoon kosher salt

½ cup whole milk

2 large eggs

1 teaspoon Bourbon Barrel Foods Aged Vanilla Extract (see Resources, page 326)

EQUIPMENT

10-inch pie pan

FOR THE CRUST: Chill the bowl, lid, and steel blade of the food processor and all of the ingredients for the crust in the freezer for 1 hour.

Preheat the oven to 350°F.

Put the flour, cocoa powder, sugar, and salt in the bowl of the food processor and process for about 1 minute to combine. Add the butter and pulse 2 or 3 times, until it is in pieces the size of a pea. Slowly add the water through the feed tube, pulsing 4 or 5 times to incorporate it.

Turn the dough out onto a lightly floured work surface and shape it into a disk. Start rolling it by lightly pressing it with the rolling pin and rolling from the center out. Do this a few times, then pick up the dough, rotate it a quarter turn, and roll again. Continue to roll, rotating the dough and flouring the work surface and the dough as needed, until you have a 12-inch circle that is ⅛ inch thick.

Loosely fold the dough into quarters and unfold it into the center of a 10-inch pie pan. Gently fit it into the bottom and up the sides of the pan. You should have an approximately ½-inch overhang around the edges. Remove the extra dough with a knife and then fold the dough under to create a clean edge around the entire rim. Let the crust rest for 10 minutes to prevent shrinkage.

Line the crust with foil or parchment paper and fill it with pie weights. Place the crust on a rimmed baking sheet and bake for 15 minutes. Remove the crust from the oven and remove the weights and foil or parchment. Prick the bottom of the crust a few times with a fork. Return it to the oven and bake it for about 7 minutes, until it is golden and appears dry. Cool completely before filling the pie. Reduce the oven temperature to 300°F.

FOR THE FILLING: Put the butter and chocolate in the top of a double boiler. Fill the bottom of the double boiler with water, insert the top, and set the double boiler over low heat; the water should never be hotter than a simmer. Stir the butter and chocolate with a silicone spatula until they melt, scraping down the sides as necessary and being careful not to incorporate air. Remove from the heat.

Combine the sugar, cocoa powder, and salt in a large bowl. Whisk in the milk, then whisk in the eggs and vanilla. Fold in the chocolate and butter mixture. Pour the filling into the chocolate crust.

Bake the pie for 30 minutes. Rotate it front to back and bake for 15 additional minutes, or until the filling is set and no longer jiggles in the center. Let the pie cool to room temperature on a baking rack, then refrigerate it for at least 2 hours before slicing. Serve at room temperature.

Tightly covered, the pie will keep in the refrigerator for up to 3 days.

BUTTERMILK PIE

WITH CORNMEAL CRUST

SERVES 8

I go bonkers over buttermilk. When I was a kid, I drank glasses of the stuff. Occasionally I would stir in some sugar to sweeten it up for a late-night treat, and this pie reminds me of that. It's important that you use really good buttermilk here. I get mine from Celeste Albers; it's the thickest and most intense buttermilk you've ever seen. When Celeste first brought her buttermilk to us, she immediately told us about her buttermilk pie recipe. My version has a cornmeal crust, which means there are at least two of my favorite things in this dessert.

CRUST

1 cup all-purpose flour, plus more for the work surface

⅓ cup cornmeal, preferably Anson Mills Antebellum Fine Yellow Cornmeal (see Resources, page 326)

½ teaspoon kosher salt

¼ cup Rendered Fresh Lard (page 316), chilled

4 tablespoons unsalted butter, cut into ½-inch pieces and chilled

¼ cup ice water

FILLING

1½ cups sugar

¼ cup all-purpose flour

½ teaspoon kosher salt

3 large eggs, at room temperature

4 tablespoons unsalted butter, melted

¼ teaspoon grated lemon zest (use a Microplane)

2 teaspoons fresh lemon juice

¾ cup whole-milk buttermilk

1 vanilla bean

EQUIPMENT

10-inch pie pan

FOR THE CRUST: Chill the bowl, lid, and steel blade of the food processor and all of the ingredients for the crust in the freezer for 1 hour.

Put the flour, cornmeal, and salt in the bowl of the food processor and pulse to combine. Add the lard and butter and pulse 2 to 3 times, until the fats are in pieces the size of a pea. Slowly add the water through the feed tube, pulsing 4 to 5 times to incorporate it.

Turn the dough out onto a lightly floured work surface and quickly gather it into a ball (you should still see spots of fat). Flatten the dough into a disk about 1 inch thick. Wrap the disk in plastic wrap and let it rest in the refrigerator for at least 1 hour, or overnight; let the dough soften a little before rolling it out if you refrigerate it overnight. *(Tightly wrapped, the dough can be frozen for up to 3 months. Thaw it in the refrigerator and let it soften a little before rolling it out.)*

Place the dough on a lightly floured work surface. Start rolling it by lightly pressing it with the rolling pin and rolling it from the center out. Do this a few times, then pick up the dough, rotate it a quarter turn, and roll again. Continue to roll, rotating the dough and flouring the work surface and the dough as needed, until you have a 12-inch circle that is ⅛ inch thick.

Loosely fold the dough into quarters and unfold it into the center of a 10-inch pie pan. Gently fit it into the bottom and up the sides of the pan. You should have an approximately ½-inch overhang around the edges. Fold the excess under and crimp the edge: with one hand on the inside of the edge and the other hand on the outside, use the index finger of your inside hand to push the dough between the thumb and index finger of your outside hand to form a V shape. Repeat all around the entire rim. Place the pie shell in the freezer while you make the filling.

Preheat the oven to 350°F.

FOR THE FILLING: Put the sugar, flour, and salt in a small bowl and mix together with a whisk. Whisk the eggs in a medium bowl until blended. Add the flour mixture, whisking to combine. Add the butter, lemon zest, and lemon juice and whisk well. Add the buttermilk and whisk well.

Split the vanilla bean lengthwise in half and, using the back of a paring knife, scrape out the seeds. Add the seeds to the filling and whisk to combine.

Place the pie pan on a rimmed baking sheet. Gently pour the filling into the shell. Bake the pie for 40 to 45 minutes, or until the custard is set and no longer jiggles in the center. Let the pie cool to room temperature on a baking rack, then refrigerate it for at least 2 hours before slicing. Serve at room temperature.

Tightly covered, the pie will keep in the refrigerator for up to 3 days.

RHUBARB BUCKLE
WITH POPPY SEED–BUTTERMILK ICE CREAM

SERVES 8 TO 10

Rhubarb is a cool-weather perennial, and a good rhubarb patch will last for years. The stalks give you a mouth-puckering tang, so it takes a mean dose of sugar to round out the stuff. Even then, you may think you've lit into a jar of alum if you haven't added enough sweetness. But I have a thing for desserts that aren't overly sweet. The sourness of the rhubarb in this dish works perfectly for me and goes well with the understated nuttiness of the poppy seed ice cream.

Note that the ice cream base must be chilled for at least 12 hours, or overnight, before freezing.

RHUBARB BUCKLE

9 tablespoons unsalted butter, at room temperature

3 cups ¼-inch-thick slices peeled rhubarb

1 cup sugar

2 cups all-purpose flour, preferably White Lily

2 teaspoons baking powder

¼ teaspoon kosher salt

½ cup whole milk

1 teaspoon pure vanilla extract

1 large egg

STREUSEL

4 tablespoons unsalted butter

¼ cup packed light brown sugar

¼ cup rolled oats

½ cup chopped pecans

¼ cup all-purpose flour, preferably White Lily

¼ teaspoon kosher salt

Poppy Seed–Buttermilk Ice Cream (recipe follows)

EQUIPMENT

One 9-inch springform pan

FOR THE RHUBARB: Preheat the oven to 375°F. Spray a 9-inch springform pan with nonstick spray and wipe out the excess with a paper towel.

Melt 1 tablespoon of the butter in a large skillet over medium heat. Add the rhubarb and cook, stirring frequently until tender, about 5 minutes. Add ¼ cup of the sugar and stir to dissolve it. Remove the skillet from the heat and allow the rhubarb to cool completely.

MEANWHILE, MAKE THE STREUSEL: Using your hands or a fork, mix the butter, brown sugar, oats, pecans, flour, and salt in a medium bowl. Set aside.

FOR THE BUCKLE: Put the flour, baking powder, and salt in a small bowl and whisk to combine. Put the milk, vanilla, and egg in another small bowl and whisk to combine.

In the bowl of a stand mixer fitted with the paddle attachment, or in a large mixing bowl, using a hand mixer, cream the remaining 8 tablespoons butter and ¾ cup sugar until fluffy, about 3 minutes. Alternately add the flour mixture and the milk, starting and ending with flour. Fold in the rhubarb. Pour the batter into the prepared pan.

Sprinkle the streusel over the top. Bake for 30 to 40 minutes, until the buckle is golden brown and a cake tester inserted into the center comes out clean. Let stand for 15 minutes. Serve the buckle warm, with the ice cream.

The buckle is best the day that it is made, but it will keep, covered, for up to 3 days at room temperature. Reheat in a 325°F oven for 7 to 10 minutes.

POPPY SEED–BUTTERMILK ICE CREAM

MAKES 1¼ QUARTS

1½ cups heavy cream

¾ cup sugar

6 large egg yolks

1½ cups whole-milk buttermilk

2 sheets gold-strength gelatin (see Resources, page 326)

½ cup poppy seeds

½ cup fresh lemon juice (from about 3 lemons)

Put the cream in a medium heavy-bottomed saucepan and heat over medium heat until it reaches 180°F on an instant-read thermometer. When you see steam coming off the cream it is nearly there. Remove the saucepan from the heat.

Put the sugar and egg yolks in a medium bowl and whisk them together until thick, fluffy, and doubled in volume. Transfer the mixture to a small heavy-bottomed saucepan and stir constantly with a silicone spatula over medium-low heat until the temperature reaches 165°F. Slowly add ½ cup of the hot cream into the egg mixture, whisking constantly. Slowly whisk the egg mixture into the saucepan of cream.

Heat the cream, sugar, and eggs over medium heat, whisking constantly, until the mixture reaches 165°F. Remove from the stove and allow to cool for 10 minutes.

While the cream mixture is cooling, put the buttermilk in a medium bowl and place the sheets of gelatin in it to soften, about 5 minutes.

Lift the softened gelatin sheets from the buttermilk, gently wring them to remove excess liquid, and add to the cream mixture. Add the buttermilk, poppy seeds, and lemon juice and stir well. Pour into a container and refrigerate; when completely cold, cover. Refrigerate for at least 12 hours, or overnight.

The next day, pour the ice cream base into an ice cream maker and freeze, following the manufacturer's instructions. Transfer the ice cream to a freezer container and freeze until you are ready to serve it.

Tightly covered, the ice cream will keep for up to 1 month in the freezer.

CAROLINA GOLD RICE PUDDING
WITH CANDIED KUMQUATS

SERVES 8

One of the things that I love about Carolina Gold rice is its aromatic nature. It's floral and sweet and a natural fit for rice pudding. I don't make my rice pudding with a ton of sugar. I pair it with some candied kumquats from the larder, but feel free to make this with any citrus fruit that you get your hands on.

CANDIED KUMQUATS

12 ounces kumquats
1 vanilla bean
½ cup sugar
1 cup water

RICE PUDDING

2 tablespoons confectioners' sugar
4 cups water
1 cup Anson Mills Carolina Gold Rice (see Resources, page 326)
7 cups whole milk
1 vanilla bean
1 cinnamon stick
1 cup granulated sugar
3 large egg yolks

FOR THE KUMQUATS: Using a sharp knife, slice the kumquats as thin as possible and place them in a medium nonreactive saucepan. Split the vanilla bean lengthwise in half and, using the back of a paring knife, scrape out the seeds (see Note). Add the seeds, pod, sugar, and water to the saucepan and bring to a boil over high heat, then reduce the heat to medium and cook, stirring occasionally, until the kumquats are translucent, 20 to 30 minutes.

Remove from the heat and allow to cool to room temperature. (In a tightly covered container, the kumquats will keep for 1 month in the refrigerator. Bring to room temperature to serve.)

FOR THE RICE PUDDING: Bring the confectioners' sugar and water to a boil in a large saucepan over high heat. Stir in the rice, bring back to a boil, and boil for 5 minutes. Drain the rice in a strainer and rinse it under cold water until the water runs clear. Drain again.

Wash and dry the saucepan. Add the rice and milk. Split the vanilla bean lengthwise in half and, using the back of a paring knife, scrape out the seeds (see Note). Add the seeds, cinnamon stick, and granulated sugar to the saucepan, bring the rice to a simmer over medium heat, and cook, stirring occasionally, until the rice is tender, about 15 minutes. Remove the saucepan from the heat and let the rice stand for 5 minutes. Remove the cinnamon stick.

Whisk the yolks into the rice one at a time. (The mixture will look too thin, but it will thicken as it cools.) Divide the rice pudding among eight serving dishes. Cover each pudding by laying a piece of plastic wrap directly against the surface, to prevent a skin from forming. Refrigerate for at least 3 hours.

To serve, bring the rice pudding to room temperature. Garnish with the candied kumquats.

The rice pudding can be refrigerated for up to 1 day.

NOTE

Save scraped vanilla beans and store them in bourbon. They will infuse the bourbon with flavor, and you can substitute it for vanilla extract.

CALAS

In New Orleans, sweetened rice fritters are sometimes known as *calas*. They shouldn't be confused with the featherlight flour beignets that migrated south with the French Acadians from Nova Scotia. Rice fritters are not very complicated, but they have a very long history. Karen Hess traced them to Persia in her iconic work *The Carolina Rice Kitchen,* which contains numerous variations on the theme, including savory versions with seafood. My version is sweet and gets a sorghum drizzle to finish.

6 cups water

1 cup Anson Mills Carolina Gold Rice (see Resources, page 326)

2 quarts peanut oil

¼ cup all-purpose flour

2 tablespoons Anson Mills Rice Flour (see Resources, page 326)

3 tablespoons granulated sugar

2 teaspoons baking powder

½ teaspoon kosher salt

2 large eggs, lightly beaten

½ teaspoon pure vanilla extract

Pinch of grated nutmeg

Confectioners' sugar for dusting

Sorghum for drizzling (see Resources, page 326)

EQUIPMENT
1-ounce scoop

Bring the water to a boil in a medium heavy-bottomed saucepan over medium-high heat. Reduce the heat to medium, add the rice, stir once, and bring to a simmer. Simmer gently, uncovered, stirring occasionally, until the rice is overcooked and a little mushy, about 15 minutes. Drain the rice.

Pour the peanut oil into a deep fryer or a large pot and heat it to 360°F. While the oil is heating, mix together the flours, granulated sugar, baking powder, and salt in a small bowl.

Mix the cooked rice, eggs, vanilla, and nutmeg together in a medium bowl, being careful not to break up the rice too much. Fold in the dry ingredients.

Use a 1-ounce scoop to portion the calas. When the oil is hot, dip the scoop into the hot oil to coat it, then fill the scoop with batter and tap it once or twice to release any air pockets. Hold the scoop 1 inch above the hot oil and carefully drop the batter into the oil. Repeat to make more fritters, without crowding the pot. Turning them in the oil to keep them cooking evenly, fry the calas until golden brown, about 3 minutes. Carefully tap the calas with a mixing spoon to loosen them if they stick to the fryer basket or the bottom of the pot. To test for doneness, gently cut one fritter in half and make sure the batter is cooked through. Using a skimmer, remove the calas from the oil to drain on paper towels. Make sure that the oil returns to 360°F before you put in the next batch.

Sprinkle the calas with confectioners' sugar and serve hot, drizzled liberally with sorghum.

SWEET POTATO DOUGHNUTS WITH BOURBON CARAMEL

MAKES 36 TO 38
DOUGHNUTS AND
DOUGHNUT HOLES

Making your own doughnuts can't be beat—you get to eat them hot right out of the fryer. Roasting the sweet potatoes first adds a caramelized undertone to their natural sweetness. Drizzle the doughnuts with some caramel sauce while they're hot. Make both doughnuts and doughnut holes—nothing goes to waste.

DOUGHNUTS

2 large sweet potatoes (about 2 pounds total), scrubbed and patted dry

2 cups whole milk

½ cup canola oil, plus 3 to 4 quarts for deep-frying

½ cup sugar

2 teaspoons active dry yeast

2 large eggs, lightly beaten

5½ cups all-purpose flour, plus more for the work surface

1½ teaspoons kosher salt

½ teaspoon baking powder

¼ plus ⅛ teaspoon baking soda

¼ teaspoon ground cinnamon

¼ teaspoon ground ginger

⅛ teaspoon ground cloves

BOURBON CARAMEL

2 cups sugar

¼ teaspoon white vinegar

½ cup warm water

1 cup heavy cream

1 ounce bourbon, preferably Buffalo Trace or Pappy Van Winkle

½ teaspoon fine sea salt

EQUIPMENT

Candy thermometer

Deep fryer

FOR THE DOUGHNUTS: Preheat the oven to 375°F.

Pierce each sweet potato several times with a fork. Place them in a baking dish, cover it with foil, and bake them for 45 minutes, or until tender. Let cool completely.

Cut the cooled sweet potatoes in half and scoop out the flesh; discard the skins. Put the flesh in a food processor and process until smooth; you need 1 cup for this recipe.

Combine the milk, ½ cup of the canola oil, and the sugar in a medium saucepan and bring to a vigorous simmer over medium-high heat. Remove the saucepan from the heat and let the mixture cool to about 110°F, about 30 minutes.

Add the yeast to the milk mixture, stirring to dissolve. Transfer to a large bowl. Add the eggs and sweet potatoes and mix well. Stir in 4½ cups of the flour and mix well. Cover the bowl, place it in a warm place, and let the dough rise for 1 hour, or until it has doubled in size.

Combine the remaining cup of flour with the salt, baking powder, baking soda, cinnamon, ginger, and cloves. Punch down the dough and knead the flour mixture into it, mixing well. Cover the bowl, place it in a warm place, and let the dough rise for 30 minutes. The dough will be loose.

MEANWHILE, FOR THE CARAMEL: Combine the sugar, vinegar, and water in a medium saucepan, stir well, and bring to a boil over medium heat; do not stir again. Heat the sugar mixture, swirling the pan gently every few minutes for even cooking, until it is a pale golden color (it will register 340°F on a candy thermometer), about 10 minutes.

Remove the saucepan from the heat and slowly whisk in the cream and then the bourbon (if you add them too fast, the caramel will boil up and you may get burned). Add the salt. The caramel will look thin, but it will thicken once cooled. Cover and set aside. *(Tightly covered, the caramel will keep for up to 1 month in the refrigerator. Rewarm it before serving.)*

TO FRY THE DOUGHNUTS: Turn the dough out onto a heavily floured work surface. Knead it for a minute or so, until it is smooth and elastic. Roll the dough out into a large ½-inch-thick rectangle. Using a 3¼-inch round cookie cutter dipped in flour, cut out the doughnuts. Using a 1¼-inch round cutter dipped in flour, cut holes from the center of the doughnuts. Transfer to baking sheets. Pat the scraps of dough together and cut out more doughnuts and holes. Refrigerate while you are heating the oil.

Pour the oil into a deep fryer and heat it to 325°F. Working in batches of about 3 doughnuts at a time, place them in the hot oil; don't crowd them. Brown, turning once, 3 to 4 minutes. Transfer to paper-towel-lined baking sheets. Be sure to return the temperature of the oil to 325°F between batches. Working in batches, fry the doughnut holes, turning once, about 2 minutes.

Serve the doughnuts warm, drizzled with caramel.

MY AUNT SHELL'S THANKSGIVING PUMPKIN ROLL

My aunt Shell can stir a mean pot, but baking is her specialty. She rarely makes this pumpkin roll, preferring to save it for special occasions, but she always serves it on Thanksgiving. I look forward to it all year.

Make sure that you select a good baking pumpkin. The round and bright orange ones grown for Halloween carving won't taste the best. Many of the best pumpkins for cooking are ugly, grown organically and probably with warts and knobby protrusions poking out. If you can't find a good one in your local market, an acorn or butternut squash makes a fine substitute.

ROLL

1 medium heirloom pumpkin, such as Winter Luxury (about 5 pounds)

¾ cup self-rising flour, preferably White Lily

2 teaspoons ground cinnamon

1 teaspoon ground ginger

½ teaspoon grated nutmeg

½ teaspoon kosher salt

3 large eggs

1 cup granulated sugar

1 teaspoon fresh lemon juice

¼ cup confectioners' sugar

FILLING

1½ cups heavy cream

¾ cup plus 2 tablespoons confectioners' sugar

1 pound cream cheese, at room temperature

1 cup finely chopped black walnuts (see Resources, page 326)

EQUIPMENT

15-by-10-inch jelly-roll pan

FOR THE ROLL: Preheat the oven to 375°F. Place a rack on a rimmed baking sheet.

Cut the top off the pumpkin. Scoop out all the seeds and strings and discard them. Put the pumpkin on the rack on the baking sheet and bake for about 30 minutes, until it is fork-tender. Allow to cool to room temperature, at least 30 minutes.

When it is cool, scoop the flesh from the pumpkin (discard the skin), transfer to a food processor, and process until smooth, about 3 minutes. You need ⅔ cup puree for this recipe. Freeze any extra for a later use, such as soup.

Spray a 15-by-10-inch jelly-roll pan with nonstick baking spray. Wipe the pan with a paper towel to remove excess spray and line it with wax paper.

Combine the flour, cinnamon, ginger, nutmeg, and salt in a small bowl. In the bowl of a stand mixer fitted with the paddle attachment, or in a large mixing bowl, using a hand mixer, beat the eggs and granulated sugar until thick. If using a stand mixer, take the bowl off the stand. Add the pumpkin and lemon juice to the bowl and stir to combine. Fold in the flour mixture.

Spread the batter evenly in the prepared jelly-roll pan. Bake for 13 to 15 minutes, until the cake springs back when touched and the sides have begun to pull away from the pan. (If using a dark pan, check for doneness at 10 minutes.)

Sprinkle a tea towel with the confectioners' sugar. Loosen the cake from the sides of the pan and turn it out onto the towel. Peel off the wax paper. Starting from a long side, roll the cake up in the tea towel. Cool it on a baking rack for 1 hour (do not put it in the refrigerator). Rolling the pumpkin roll too tightly or cooling it for too long can cause serious breaks, but there will be some small cracks regardless.

FOR THE FILLING: Combine the cream and 2 tablespoons of the confectioners' sugar in the bowl of a stand mixer fitted with the whisk attachment, or use a medium mixing bowl and a hand mixer. Whip on low speed, increasing the speed as the cream thickens. It is ready when it holds stiff peaks. If using a stand mixer, transfer the cream to another bowl.

Put the cream cheese and the remaining ¾ cup confectioners' sugar in the bowl of the stand mixer and use the paddle attachment, or use a large mixing bowl and a hand mixer. Beat the cream cheese and sugar on medium speed until smooth, about 1 minute. Add the whipped cream and walnuts and beat on low speed until combined, about 10 seconds.

Unroll the cake. Place it on a piece of parchment. Using an offset spatula, spread the filling over the roll, leaving a ¼-inch border around the edges. Using the parchment to assist, again starting from a long side, tightly roll the cake up. Refrigerate for at least 1 hour.

TO SERVE, cut into slices. Tightly wrapped in plastic wrap, the roll will keep for 3 days in the refrigerator.

CHEWY BENTON'S BACON CARAMELS

MAKES EIGHTY-ONE 1-INCH PIECES

Allan Benton is one of my heroes. These days he has a cult following, and rightfully so. His bacon is amazing, and no other bacon tastes like it. So if you want to make these bacon caramels to taste just like mine do, look him up and buy yourself a package of his bacon through mail order. The intense hickory smoke of Allan's cured pork is the secret complement to the caramel in this classic chew, and that smokehouse flavor lurks in the background, haunting you until you go back for another piece.

4 ounces Benton's bacon (see Resources, page 326), finely minced

¾ cup heavy cream

½ vanilla bean

1 cup sugar

¾ cup sorghum (see Resources, page 326)

4 tablespoons unsalted butter

Maldon or other flaky sea salt

EQUIPMENT

Candy thermometer

Line a 9-inch square pan with heavy-duty aluminum foil, leaving an overhang on two opposite sides to use as handles to lift the caramel out. Lightly spray the foil with nonstick baking spray.

Put the bacon in a skillet large enough to hold it in one layer and cook it over medium-low heat, stirring frequently, until the fat is rendered and the bits of bacon are crispy, 4 to 5 minutes. Transfer the bacon to a paper towel to drain, reserving the fat. You need ¼ cup of bacon fat for this recipe. Reserve the bacon for another use, such as on a salad or pasta.

Combine the cream and bacon fat in a medium heavy-bottomed saucepan and bring to a boil over medium heat. Remove the pan from the heat.

Cut the vanilla bean lengthwise in half and, using the back of a paring knife, scrape out the seeds. Add the seeds to the cream, stir to combine, and cover the pan to keep warm.

Combine the sugar and sorghum in a medium saucepan and stir over medium heat to dissolve the sugar. Then attach a candy thermometer to the pan and cook until the temperature reaches 312°F. Remove the saucepan from the heat and carefully add the cream mixture; once the intense bubbling has subsided, stir to combine. Return the saucepan to the heat and cook the caramel until the temperature reaches 250°F. Remove from the heat and stir in the butter.

Pour the mixture into the prepared pan. Refrigerate until firm, about 20 minutes.

Sprinkle the caramel lightly with sea salt. Lift the caramel out of the pan, using the two overhanging ends of aluminum foil. Cut into 1-inch squares with a long chef's knife.

To wrap the caramels, cut eighty-one 3-inch squares of wax paper. Place a caramel in the middle of each square, fold one side over the caramel, then fold over the opposite side, and twist the ends to secure.

The caramels can be stored in an airtight container for up to 3 days.

MY GRANDMOTHER'S HILLBILLY BLACK WALNUT FUDGE

MAKES EIGHTY-EIGHT 1-INCH PIECES

In my family, fudge is a holiday dessert. And it may come as a shock to some, but the key ingredient in this fudge is Velveeta cheese. The ultracreamy nature of the processed whey melts more evenly than traditional cheese, and it presented twentieth-century home cooks with a new texture when it was first introduced. Well, everyone knows I am dedicated to heirloom ingredients; now I suppose you can add Velveeta to the list.

1 pound Velveeta cheese, cut into ½-inch-thick slices

1 pound unsalted butter, cut into ½-inch-thick slices

Four 1-pound boxes confectioners' sugar

1 cup unsweetened cocoa powder

1 cup chopped black walnuts (see Resources, page 326)

1 teaspoon Bourbon Barrel Foods Aged Vanilla Extract (see Resources, page 326)

Spray a 9-by-13-inch pan lightly with nonstick baking spray.

Put the Velveeta and butter in the top of a double boiler. Fill the bottom of the double boiler with water, insert the top, and set the double boiler over low heat; the water should never be hotter than a simmer. Stir the Velveeta and butter together with a silicone spatula until melted and combined, scraping down the sides as necessary, about 8 minutes. Transfer the mixture to a large bowl and set aside.

Put the confectioners' sugar and cocoa in a large bowl and whisk together, making sure that no lumps remain. Add the nuts and stir to combine.

Add the sugar mixture to the cheese mixture, then add the vanilla and stir until the sugar is dissolved and the mixture is smooth. Pour the fudge into the prepared pan. Tap the pan on the counter to remove any air bubbles and smooth the top with a small offset spatula. Refrigerate for at least 8 hours; wait until the fudge is cold before covering it, so that moisture won't form on the top. Cut the fudge into 1-inch squares. Serve at room temperature.

Tightly covered, the fudge will keep for up to 1 week in the refrigerator. Tightly wrapped, it can be frozen for up to 3 months. Thaw it in the refrigerator and bring to room temperature before serving.

THE BASICS

THE BASICS

I consider "basics" to be the routine but not insignificant recipes that I keep stored away in my back pocket. These are formulas that I have been tweaking and improving on for years, and now they are as good as it gets. I use these recipes frequently; they function as a foundation for many other dishes. View them as the backbone of your cooking or as an ace of spades tucked into your shirtsleeve, but keep in mind that the recipes are merely a starting point. Feel free to play with them or adapt them to your personal taste and cuisine.

BASIC MEAT SAUCE

MAKES 1½ CUPS

1 tablespoon canola oil

1¾ cups (6½ ounces) thinly
 sliced shallots

1 cup (4 ounces) chopped
 peeled carrots

½ cup (2 ounces) chopped celery

1 garlic clove, smashed and peeled

1 cup red wine, preferably
 Cabernet Sauvignon

1 cup dry Madeira

4 cups Veal or Beef Stock (opposite)

2 cups Chicken Stock (page 318)

1 bouquet garni (a 4-inch piece of
 leek green, 4 thyme sprigs,
 2 fresh bay leaves, and
 6 black peppercorns, wrapped
 in cheesecloth and tied tightly
 with butcher's twine)

1 teaspoon fresh lemon juice

1 teaspoon soy sauce

½ teaspoon sherry vinegar

1 tablespoon unsalted butter

1 teaspoon kosher salt

Heat the canola oil in a large saucepan over high heat until starting to smoke. Add the shallots, carrots, celery, and garlic and cook for 1 minute without stirring. Then continue to cook the vegetables, stirring constantly, until darkly caramelized, almost black in color, about 6 minutes.

Reduce the heat to medium-high, add the wine and Madeira, and bring to a boil, then boil until they have almost evaporated, about 12 minutes. Reduce the heat to low, add the veal stock, chicken stock, and bouquet garni and bring to a simmer, then simmer to reduce the liquid until the sauce thickens enough to coat the back of a spoon, about 1¾ hours. Strain the sauce through a fine-mesh strainer into a small saucepan. *(The sauce can be made to this point, cooled, and refrigerated, tightly covered, for up to 1 week.)*

To finish the sauce, heat over medium heat until hot, about 3 minutes. Add the lemon juice, soy sauce, and sherry vinegar. Remove the saucepan from the stove and swirl in the butter and salt. Use at once.

RAMP TARTAR SAUCE

MAKES 2 CUPS

An amped-up tartar sauce with pickled ramps.

1¾ cups mayonnaise, preferably
 Duke's (see Resources, page 326)

3 tablespoons ⅛-inch diced white
 Pickled Ramp bulbs (page 233), plus
 2 tablespoons brine from the ramps

2 tablespoons plus 1 teaspoon
 chopped drained capers

2 tablespoons ⅛-inch dice shallots

1 heaping tablespoon chopped dill

1 tablespoon fresh lemon juice

Put all the ingredients in a small bowl and stir to combine. Cover and refrigerate until chilled before serving.

Tightly covered, the tartar sauce will keep for up to 4 days in the refrigerator.

BLACK TRUFFLE JUS

1 tablespoon canola oil

3 shallots, thinly sliced

1 large carrot, chopped

2 celery ribs, chopped

1 garlic clove, smashed and peeled

1 cup red wine, preferably Cabernet Sauvignon

1 cup dry Madeira

4 cups Veal Stock (page 320)

2 cups Chicken Stock
  (page 318)

1 bouquet garni (1 leek top, 4 thyme sprigs,
  2 fresh bay leaves, and 6 whole black peppercorns,
  wrapped in cheesecloth and tied tightly with
  butcher's twine)

2 tablespoons truffle juice
  (see Resources, page 326)

1 tablespoon plus 1 teaspoon minced black truffle

1 teaspoon fresh lemon juice

1 teaspoon soy sauce

½ teaspoon red wine vinegar, preferably Banyuls
  (see Resources, page 326)

1 tablespoon unsalted butter

Heat the canola oil in a large saucepan over high heat until it is start-ing to smoke. Add the shallots, carrot, celery, and garlic and cook for 1 minute, without stirring. Then continue to cook the vegetables, stir-ring constantly, until they are darkly caramelized and almost black, about 6 minutes.

Add the wine and Madeira, decrease the heat to medium-high, and cook until the liquid has almost completely evaporated, about 6 minutes. Reduce the heat to low, add the veal stock, chicken stock, and bouquet garni, and simmer until the sauce coats the back of a spoon, about 1½ hours. Strain the sauce through a fine-mesh strainer into a container. Cover and refrigerate. Tightly covered, the jus will keep for up to 2 days in the refrigerator.

VADOUVAN JUS

1 cup dry white wine

1 small shallot, shaved as thin as possible

1 garlic clove

2 cups heavy cream

2 tablespoons Vadouvan Spice (page 309)

Put the wine, shallot, and garlic in a medium saucepan, bring to a boil over high heat, and cook until the liquid has almost completely evaporated, about 7 minutes. Add the cream and vadouvan and bring to a simmer, then reduce the heat to low and simmer until the mixture coats the back of a spoon, about 10 minutes.

Strain the jus into a container, cool to room temperature, cover, and refrigerate. Tightly covered, the jus will keep for up to 2 days in the refrigerator.

CONVERSION CHARTS

Here are rounded-off equivalents between the metric system and the traditional systems used in the United States to measure weight and volume.

WEIGHTS		VOLUME			OVEN TEMPERATURE			
US/UK	METRIC	AMERICAN	IMPERIAL	METRIC		°F	°C	GAS MARK
¼ OZ	7 G	¼ TSP		1.25 ML	VERY COOL	250–275	130–140	
½ OZ	15 G	½ TSP		2.5 ML				½–1
1 OZ	30 G	1 TSP		5 ML	COOL	300	148	2
2 OZ	55 G	½ TBSP (1½ TSP)		7.5 ML	WARM	325	163	3
3 OZ	85 G	1 TBSP (3 TSP)		15 ML	MEDIUM	350	177	4
4 OZ	115 G	¼ CUP (4 TBSP)	2 FL OZ	60 ML	MEDIUM HOT	375–400	190–204	5–6
5 OZ	140 G	⅓ CUP (5 TBSP)	2½ FL OZ	75 ML	HOT	425	218	7
6 OZ	170 G	½ CUP (8 TBSP)	4 FL OZ	125 ML	VERY HOT	450–475	232–245	8–9
7 OZ	200 G	⅔ CUP (10 TBSP)	5 FL OZ	150 ML				
8 OZ (½ LB)	225 G	¾ CUP (12 TBSP)	6 FL OZ	175 ML				
9 OZ	255 G	1 CUP (16 TBSP)	8 FL OZ	250 ML				
10 OZ	285 G	1¼ CUPS	10 FL OZ	300 ML				
11 OZ	310 G	1½ CUPS	12 FL OZ	350 ML				
12 OZ	340 G	1 PINT (2 CUPS)	16 FL OZ	500 ML				
13 OZ	370 G	2½ CUPS	20 FL OZ (1 PINT)	625 ML				
14 OZ	400 G	5 CUPS	40 FL OZ (1 QT)	1.25 L				
15 OZ	425 G							
16 OZ (1 LB)	450 G							

Published by Artisan

A division of Workman Publishing Company, Inc.

225 Varick Street

New York, NY 10014-4381

artisanbooks.com

Published simultaneously in Canada by Thomas Allen & Son, Limited.

Library of Congress Cataloging-in-Publication Data

Brock, Sean.
   Heritage / Sean Brock.
     pages cm
   Includes index.
   ISBN 978-1-57965-463-4
   1. Cooking, American—Southern style. I. Title.
   TX715.2.S68B75 2014
   641.5975—dc23   2014005022

Design by Michelle Ishay-Cohen

Printed in China

First printing, September 2014

10 9 8 7 6 5 4 3 2 1